CW00522518

"Dr. Diallo presents a rich analysis of the emergence and evolution of bauxite mining in Guinea, bringing a strong community lens to themes of stability and equality, at one of the critical starting points for the global aluminium supply chain".

– Dr Fiona Solomon, Chief Executive Officer,
Aluminium Stewardship Initiative

"This is a brilliant study of the politics and intrigues surrounding Bauxite mining in Guinea, and the best I have read on the subject. It is well-written and brilliantly presented. This book genuinely breaks new grounds, and academics and practitioners alike will benefit from the rich data and analysis. The author has indeed made a significant contribution to the politics of natural resource management in a developing society and we are totally in her debt for this".

– Dr Abiodun Alao, Professor of African Studies and
Director, African Leadership Centre, King's College London, UK

"Going beyond the negative perception of mining in Africa, this book provides an overview of the exploitation of bauxite in Guinea by highlighting the interactions of key stakeholders. The data presented forms an important source of information and inspiration for national reforms focused on good governance and on concerns of grassroots populations".

– Dr Moustapha Keïta-Diop, Professor and Dean of the
Faculty of Social Sciences, l'Université Général Lansana
Conté de Sonfonia, Conakry, Republic of Guinea

"Penda's analysis helps better understand a complex period in Guinea. I expect her insights will be more widely applicable!"

– Nic Clift, Senior Industry Fellow, RMIT University,
Australia; Former Director General of Compagnie
des Bauxites de Guinee (CBG) 2004–2007

Regime Stability, Social Insecurity and Bauxite Mining in Guinea

This book explores how bauxite mining has affected local and national political dynamics in Guinea over the past 55 years, providing an overview of mining interactions with social, economic and political spheres.

Guinea is amongst the world's top producers of bauxite, and the country's rich mineral presence has numerous implications on local communities and national policy. Guinea is an interesting and highly relevant case study in assessing the impact of bauxite mining on regime stability and social insecurity. The author offers a clear understanding of the role of mining during the Touré and Conté regimes and analyses how changes since the election of Condé in 2010 have affected the socio-political and economic development of Guinea. The author also offers analysis on how bauxite mining has led to the emergence of new forms of social contracts, sustained by mining companies instead of the state. Finally, the book argues that understanding the stabilising and destabilising potential of mining is key to ensuring long-term, sustainable, stable and inclusive growth of mineral-resource-rich countries. The book concludes by highlighting the relevance of the findings in Guinea for the wider African extractives sector.

The book will be of interest to a wide range of scholars, including those working in the areas of African studies, political science, political economy, sustainable development and corporate social responsibility. The book will be relevant for academics, business actors, NGOs, policy-makers and students interested in the African mining sector.

Penda Diallo is Lecturer in Sustainable Mining, Camborne School of Mines, University of Exeter, UK.

Routledge Studies of the Extractive Industries and Sustainable Development

Mining in Latin America
Critical Approaches to the New Extraction
Edited by Kalowatie Deonandan and Michael L. Dougherty

Industrialising Rural India
Land, Policy, Resistance
Edited by Kenneth Bo Nielsen and Patrik Oskarsson

Governance in the Extractive Industries
Power, Cultural Politics and Regulation
Edited by Lori Leonard and Siba Grovogui

Social Terrains of Mine Closure in the Philippines
Minerva Chaloping-March

Mining and Sustainable Development
Current Issues
Edited by Sumit K. Lodhia

Africa's Mineral Fortune
The Science and Politics of Mining and Sustainable Development
Edited by Saleem Ali, Kathryn Sturman and Nina Collins

Energy, Resource Extraction and Society
Impacts and Contested Futures
Edited by Anna Szolucha

Regime Stability, Social Insecurity and Bauxite Mining in Guinea
Developments Since the Mid-Twentieth Century
Penda Diallo

For more information about this series, please visit: www.routledge.com/series/REISD

Regime Stability, Social Insecurity and Bauxite Mining in Guinea

Developments Since the
Mid-Twentieth Century

Penda Diallo

LONDON AND NEW YORK

First published 2020
by Routledge
2 Park Square, Milton Park, Abingdon, Oxon OX14 4RN

and by Routledge
52 Vanderbilt Avenue, New York, NY 10017

Routledge is an imprint of the Taylor & Francis Group, an informa business

First issued in paperback 2021

British Library Cataloguing-in-Publication Data
A catalogue record for this book is available from the British Library

Library of Congress Cataloging-in-Publication Data
Names: Diallo, Penda, author.
Title: Regime stability, social insecurity and bauxite mining in Guinea : developments since the mid-twentieth century / Penda Diallo.
Description: Abingdon, Oxon ; New York, NY : Routledge, 2020. | Includes bibliographical references and index.
Identifiers: LCCN 2019034535 (print) | LCCN 2019034536 (ebook) | ISBN 9780367252113 (hardback) | ISBN 9780429286544 (ebook)
Subjects: LCSH: Aluminum mines and mining—Guinea. | Mineral industries—Government policy—Guinea. | Guinea—Economic conditions—1958–1984. | Guinea—Economic conditions—1984–
Classification: LCC HD9539.A63 G8535 2020 (print) | LCC HD9539.A63 (ebook) | DDC 333.854926096652—dc23
LC record available at https://lccn.loc.gov/2019034535
LC ebook record available at https://lccn.loc.gov/2019034536

ISBN: 978-0-367-25211-3 (hbk)
ISBN: 978-1-03-208749-8 (pbk)
ISBN: 978-0-429-28654-4 (ebk)

Typeset in Times New Roman
by Apex CoVantage, LLC

To my father and mother, for being a constant source of inspiration, for the values they taught me and for their continuous support and blessings. To Prof. 'Funmi Olonisakin, for her inspiration and support. To Grandma and Grandpa Olonisakin, for their unconditional love. And to the vibrant Guinean youth, in their struggle for a better future.

Contents

Acknowledgments

Although I am responsible for the content of this book, many people have contributed to making it possible, and I cannot thank them enough for their support, encouragement and guidance.

I am profoundly grateful to the Guinean Ministry of Mines and Geology and their staff. Without the permission received from the Ministry of Mines and Geology, it would have been impossible for me to access data for this manuscript. I am also grateful to the Compagnie des Bauxite de Guinée and its different Managing Directors for giving me access to CBG and its resources from 2013 to 2019. Staff from both the Ministry of Mines and CBG spent extensive time with me throughout my research, and I am deeply grateful for their time. The Central Bank of Guinea also provided me with valuable data, and the time I spent with some of their staff has been crucial for my research.

I am incredibly grateful to informants in Conakry, Kamsar, Sangaredi and Boké for their openness, great hospitality and support. Many informants went out of their way to make time for me, invite me to their homes and include me in activities with their friends and family. Without the trust and openness of the different interviewees, I would not have been able to capture the implications of bauxite mining in Guinea.

Many actors were key in deepening my understanding of mining, stability and politics in Guinea. In this regard, I am grateful to all the staff from mining companies, youth groups, civil servants, NGO actors, academics, journalists, World Bank staff and diplomats who shared their views and insights with me.

The Camborne School of Mines gave me the time to write the manuscript, and I am grateful both to my department and to my colleagues, who shared their input with me. The research for this manuscript started at the University of Edinburgh, and my special thanks go to Dr Wolfgang Zeller, Prof. Paul Nugent and Dr Andrew Lawrence from the University of Edinburgh for their support, guidance, availability and productive feedback.

I received great insight from institutions including EITI (Extractive Industry Transparency Initiative) Guinea office, Publish What You Pay (PWYP) Guinea, the offices of the World Bank in Conakry and the Open Society Initiative for West Africa (OSIWA), the British Embassy in Guinea, the Aluminium Stewardship Initiative, youth in Kamsar and Boké, and academic colleagues at King's College London. The different exchanges with representatives from these institutions were essential in my understanding of the impact of bauxite mining on the Guinean society and the impact of mineral extraction in Africa.

During my archival research in Paris and Nantes, family and friends hosted me, and I am grateful to them for that.

I would also like to thank friends, colleagues and family members who set aside time to review the chapters of the book and who shared their input and feedback with me. Often, they brought to my attention suggestions and additions that would not have occurred to me.

During my time in Guinea in 2013 and 2014, before the Internet became more accessible in Guinea, there were companies who made their Conakry offices available to me. I thank these companies for their support, which gave me the occasional luxury to use the Internet and have access to electricity when I needed these resources.

With considerable regret, I have decided not to mention the names of numerous individuals who assisted me in the writing of this book. The reader will notice that I identify only academics who assisted me during my PhD. The non-inclusion of individual names is to avoid painting them with a brush of association, especially because the sensitive positions of some of them caution against mentioning them by name. I apologise for this and do hope they will accept my general expression of appreciation. I would not want to put any individual at risk for what they may not even agree with in the book.

My family, extended families and friends around the world have continuously supported me during my research, and I am forever grateful and indebted to them for their support and patience.

My special thanks goes to the Routledge team for their editorial support and the publication of the manuscript.

While I am deeply grateful for all the support I received, I accept sole responsibility for any mistakes in the book.

Map of Guinea

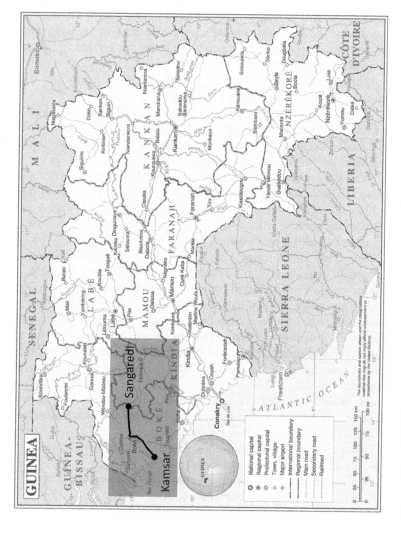

Source: Guinea, Map No. 4164 Rev.4, August 2014, United Nations

1 Introduction

1.1 A political-economic history of Guinea

1.1.1 Study background

From the 1990s until 2008, there were varying degrees of violent conflict linked to mineral resources in most of the countries bordering Guinea. In 2004, Sierra Leone and Liberia emerged from over a decade of armed conflict linked to mineral resource governance (1991–2002 for Sierra Leone; 1989–2003 for Liberia). Ivory Coast suffered from political instability because of several mutinies (1999, 2000 and 2002) and a civil war (2002–2007).

During the first 50 years of Guinea's post-independence existence, presidents Ahmed Sékou Touré (1958–1984) and Lansana Conté (1984–2008) led the country. Since 2010, Alpha Condé has led the country. Although Guinea was the victim of two major mutinies (1996 and 2008) and rebel attacks from Liberia (2000–2001), on all three occasions the country managed to avoid large-scale resource-related conflict. Despite a long record of poor governance and dictatorship during that period, Guinea managed to maintain the stability of its governing regime ('regime stability').

In the last two decades, the subject of mineral resource governance, particularly in countries with an abundance of natural resources, has been at the forefront of academic and policy debates. This is due to two factors: the ever-growing global demand for natural resources and the continued emergence of conflicts involving natural resources, usually with wider regional implications. As a result, a growing literature has emerged focused on the 'resource curse'. Some authors suggest that low economic growth, widespread poverty, political dictatorship, armed conflict or a combination of these (Collier, 2000; Bannon & Coller, 2003; Ross, 1999) will often affect developing countries that depend heavily on natural resources. Such situations make it easy for rebels or other (armed/organised) groups to mobilise rural populations, particularly with a promise of a better share of resources in the future. Such promises encouraged conflicts in Sierra Leone and Liberia.

The similarity of Guinea to countries in the region that have been affected by armed conflict and the fact that the country shows symptoms identified in the resource curse literature raise key questions: what is it that has enabled the coexistence of regime stability (the absence of change in the country's leadership) and 'social insecurity' (the lack of socio-economic development for the broader population) in Guinea? Why has Guinea not (yet) descended into armed conflict? Have mineral resources and their governance in some way accommodated the coexistence of both regime stability and social insecurity, preventing the nation from collapsing into civil war?

In this book, I make two key arguments. First, the use of mineral resources as leverage for consolidating dictatorship and authoritarian rule is central to understanding both regime stability and social insecurity in Guinea. Second, revenues provided by the bauxite mining industry and the emergence of new forms of social contract in bauxite mining regions played a crucial role in maintaining regime stability. This provided the state with a stable source of revenues, decreased the civic pressure on regimes to cater to communities in mining areas and created direct pressure on mining companies to contribute to the local livelihoods and infrastructure, with limited and often no regime support.

In this book, I consider regime stability to involve a) governmental longevity, b) the absence of structural state change, c) possession by the state of coercive powers and its use of them to minimise any threat to its longevity and d) the absence of civil war (Ake, 1975; Hurwitz, 1973; Andersen et al., 2014). Regime stability in this sense is the ability of the regime in power to maintain itself using one or a combination of the following: coercion, oppression, threats, fear, intimidation, dictatorship and manipulations of the minds of people. In the case of Guinea, I refer to regime stability as the ability of the Touré and Conté regimes to maintain power (for 26 years and 24 years respectively). Both regimes were dictatorial, yet their presidents managed to stay in power until their deaths because of the regular use of violence and threats to ensure the population obeyed their laws, regulations and personal wishes.

In this book, I define social insecurity as arising from a lack of socio-economic development. Social insecurity is taken to involve poor socio-economic development, including a) low living standards, b) increasing poverty rates, c) poor development of social infrastructure and d) unemployment. Social insecurity, as used in this book, involves the unavailability or unequal/unreliable distribution of social goods, resulting in an increasing rate of poverty across the country – as seen both under Touré and Conté and now under Condé.

Although regime stability has been maintained, living conditions for ordinary Guineans continue to be far from perfect or even satisfactory. While some West African countries where natural resources have had an important role in state collapse and civil war (Sierra Leone and Liberia) and weakness (Nigeria) have received great scholarly attention, in-depth empirical work remains rare on countries where a semblance of order still exists. In particular, there has been no detailed study to date of the dynamics of governance, regime stability, social insecurity and the politics of mineral resource governance in Guinea. The literature on mining and conflict/governance in Africa – such as that of Liberia, Sierra Leone and Nigeria – has largely omitted Guinea. So far there is no study that assesses how the emergence of different forms of social contracts has enabled regimes to maintain stability in the face of social insecurity in mineral-rich countries. Select studies have focused on bauxite mining and its governance in Guinea (Campbell, 1983, 1995, 2009, 2010; Knierzinger, 2018), but to date no study has explored how bauxite mining has influenced the emergence of new forms of social contracts. It does appear, however, that a particular set of circumstances and stakeholder responses have enabled Guinea to avoid a slide into armed conflict through the emergence of different forms of social contracts. For these reasons, academic research on the governance and politics of mineral resource extraction still suffers significant theoretical and empirical gaps, and the example of Guinea has been largely neglected.

1.1.2 *1958 to 2008: Guinea's political and economic coming of age*

I focus on the period from 1958 to 2008 for six key reasons. First, it includes political phases during which Guinea managed to avoid resource-related conflict despite the country's extremely poor record of governance, its increasing poverty and the coercive nature of its regimes. Second, during this period there was extreme resource-related violence in Sierra Leone and Liberia, which border Guinea; yet – despite external threats from rebel insurgencies in 2000 from Liberia and Sierra Leone – Guinea managed to maintain the status quo. Third, 50 years of mining has provided people the time to observe and compare the impact of mining on politics in Guinea. Fourth, this is also sufficient time to assess the economic, social and political consequences of resource mining. Fifth, throughout this period, bauxite mining remained an important source of revenue for Guinea. Finally, this period covers past regimes, not the current one, and both presidents from that period are dead; thus, it was hoped that people might be more frank and honest in their contributions to this research without fear of reprisal. Chapter 6 of this book does offer a brief overview of the governance of bauxite mining in Guinea since 2010. While this period is beyond the focus of this book, Chapter 6 serves as an entry point for those interested in undertaking further study of bauxite mining and governance in Guinea.

The book draws on Soares de Oliviera's (2007) analytical framework of the 'successful failed state', as well as discussions of different authors on the 'resource curse' and the social contract paradigms in Africa. The research also draws on Nugent's (2010) concept of social contracts, distinguishing between coercive and permissive contracts as a key analytical framework for the book. Social contracts refer to the relationship between the state and the society and, more specifically, to the relationship between the rulers and those whom they rule. Nugent (2010) identifies four types of social contracts, namely coercive, productive, liberational and permissive contracts. These different types of social contracts are discussed further in Chapter 2.

It is hoped this book will encourage more studies on both the stabilising and destabilising potential of mining, as opposed to the current focus on the negative aspect of mineral resource extraction in Africa. Guinea has some of the world's largest reserves of important resources, including bauxite, iron ore, diamonds and gold. Although Guinea has managed to maintain the coexistence of regime stability and 'social insecurity', with the increase in mining activities and the increasing number of investors, it will become crucial for Guinea to understand both the stabilising and destabilising potentials of its mineral resources to avoid future conflict. Findings from this book provide a valuable source of information for the effective reform of future policies and activities in the large-scale mining sector in Guinea and the subregion.

1.1.3   *Bauxite: the making of a nation*

Bauxite was discovered in Guinea in 1819 by a French explorer, but it was not until 1920 that exploration began in the Boké Region and continued until 1958 (Touré, 2013). By the time Guinea achieved independence, extensive research and geological surveys had been carried out to estimate bauxite reserves, enabling Guinea to survive its first years of independence and contributing to the longevity of the Touré and Conté regimes, as will be discussed in Chapters 3 and 4. Bauxite mining offered job opportunities to Guineans in the early days of independence and provided a direct and sustainable contribution to state revenues.

Guinea is one of the world's three largest producers of bauxite, yet it remains amongst the world's poorest countries, ranking 183rd out of 195 countries (UNDP, 2018). Guinea also has other significant mineral resources, including iron ore, diamonds and gold (MMG, 2018), creating significant commercial interest globally. In 2017, Guinea produced 45 million tons of bauxite – by comparison, Australia produced 83 million tons; China, 68 million tons

(USGS, 2018). Sources note that the Guinean government "expects this production to reach at least 60 million tons by 2020" (MMG, 2018:20).

Mineral extraction has dominated the Guinean economy since 1958. Over half of foreign investment in the country is from the mining sector, and it constitutes the majority of export revenues. Over the years, dependence on the mining industry has increased. In 2005, it was stated that mining accounted for about 80 per cent of the country's export revenues and 20 per cent of GDP (Huijbregts & Palut, 2005). In 2016, mining represented about 90 per cent of Guinea's export revenues and 20–25 per cent of state revenues; it corresponded to 25 per cent of the country's GDP (PAGSEM, 2016:13). Three main resources, bauxite, gold and diamonds, have been key contributors to the Guinean extractive sector, with bauxite production being by far the highest contributor to the country's mining industry.

From 1958 to 2008, economic growth in Guinea can be credited to the mining sector, which was the biggest employer after the state, the largest contributor to the state's export revenue and the most stable source of tax revenue. Industrial mining activities provided about 22,000 direct full-time jobs and created over 50,000 indirect jobs (IMF, 2008:92). Even the artisanal mining of diamonds was estimated to equate to about 103,000 jobs (Chirico et al., 2014). The formal mining sector also made a direct contribution to customs taxes, mining taxes and local taxes. Mining revenues contributed 15.8 per cent of public expenses in the areas of health, education, water and infrastructure development (MEF, 2002). As a result, it ultimately contributed to the reduction of poverty in Guinea.

Mining played an important role in the formation of the Guinean state. At the time of independence, the business transaction between Touré and the Soviet Union were based on bauxite. The bauxite production of SBK (Société des Bauxites de Kindia) was used to pay back loans and services received from the Soviet Union at independence (cf. Chapter 3). It was only in 2001 that the state actually started to account for revenues generated by the SBK. Until 2008, mining was the single most important contributor to the Guinean economy as a whole and, thus, the single most important industry for state revenue generation (BCRG, 2015; IMF, 2004, 2008). Since bauxite mining dominates the industrial mining industry, it is one of the major contributors to state revenues, thus making it an important industry for the Guinean economy.

Over the years, mining revenues made important contributions to state revenues. In 2008, bauxite accounted for about USD$596 million, or 40 per cent, of the country's total exports (United Nations Statistics Division 2009 in USGS, 2011). Bauxite is used to produce aluminium, which is used in industries such as transport, construction, engineering and packaging. It provides Guinea with important international recognition. Guinea has the world's largest reserves of high-quality bauxite, which is estimated at 7.8

billion tons (USGS, 2018). With 26 per cent of the world's total bauxite reserves, Guinea will remain an important player in the bauxite mining industry (2018, p. 2). In 2012, Guinea was the world's sixth-largest producer and exporter of bauxite; as of the writing of this book, it is now the world's third-largest producer (KPMG, 2014; USGS, 2018). This demonstrates a relatively sharp increase of bauxite mining production in Guinea over the past seven years. The Guinean government intends to continue moving up the list of top global producers; thus, to meet this goal, it has, over the last five years, attracted major companies to invest in its bauxite mining industry.

Industrial mining in Guinea started in the early 1960s, in Fria, with bauxite mining and its transformation into alumina via refining; bauxite mining then also developed in Boké-Sangarédi and Kindia-Débélé in 1973–74. From 1958 to 2008, three main bauxite mining companies were active in Guinea, and these are listed in Table 1.1.

Table 1.1 List of companies that have undertaken bauxite extraction in Guinea (1958–2008)

Companies and Shareholders	*Mineral*
Alumina Company of Guinea (ACG) since 2000; previously operated by FRIGUIA and before then it was FRIA	bauxite and alumina
☐ Initial concession signed in 1958 and resulted in a company named FRIA.	
☐ In 1973 FRIA became FRIGUIA (49% held by the Guinean government and 51% by FRIALCO – a consortium of foreign investors).	
☐ In 2000, the company was leased to ACG for 25 years.	
☐ ACG was 100% held by Russki Alumini (Russia).	
CBK (Compagnie des Bauxites de Kindia) previously known as Société des Bauxites de Kindia (SBK); before that it was Office des Bauxites de Kindia (OBK)	bauxite
☐ Concession signed in 1969.	
☐ Owned by Guinean government.	
☐ 85% controlled by Russki Alumini (Russia).	
Compagnie des Bauxites de Guinée (CBG)	bauxite
☐ Concession signed in 1963.	
☐ CBG is 49% owned by the Republic of Guinea and 51% by Halco Mining. The Halco Group is 45% owned by Alcoa Inc, 45% owned by Rio Tinto Alcan Inc and 10% by Dadco.[1]	
Parent company or affiliate out of Guinea and their headquarters: Alcoa (USA), Rio Tinto Alcan (Canada, Australia), Dadco.	

Source: MMG (2005); CTRTCM (2015); Labbé, 2005. Table compiled by author

Although Guinea possesses a wide range of mineral resources, to date beside from bauxite mining, no other major mineral extraction projects have started in Guinea. Infrastructure remains a major issue, and this is certainly one of the reasons why all the major mining projects that have been developed in Guinea to date have related to the bauxite industry. The majority of past and current bauxite mining activities are for resources that are relatively close to the coastline. The infrastructure requirements and associated costs to access shipping are thus relatively reduced. However, the increasing development of bauxite mining on the Guinean coast, particularly since 2010, has created new social, economic and environmental challenges for local communities that will need further attention.

Mining developments in Guinea appear to be coming online relatively slowly. While numerous mineral concessions have been granted to large international mining companies, by 2008 – 20 years after some of these concessions were granted – most companies had not yet started mining activities. As a result, between 1958 and 2008, only a few companies with mining licenses were actively engaged in mineral extraction in Guinea.

1.1.4   *Expectations unmet*

In 2008, after 50 years of development of the extractive industry, Guineans as a whole felt that they were yet to benefit significantly from the revenues of its mining industry. From 2004 to 2008, during national protests, citizens showed their frustration at the fact that Guinea had a wealth of resources but was faced with increasing poverty and little or no infrastructure to support development except those related to bauxite mining (ICG, 2007;). The population was angry and felt not only that taxes from bauxite mining did not contribute to the development of other regions of Guinea but also that development in the bauxite mining regions was insufficient. Since Guinea's independence, the mining sector served primarily to promote and support the political agendas of Guinea's leaders (see further discussions in Chapters 3 and 4). While this created tension within the population, in the period studied here (and during subsequent protests from 2008 to 2010), the regimes managed to prevent large-scale armed conflict.

The death of President Conté in December 2008, the seizure of power by a military junta a few hours afterwards and the violent repression of a peaceful protest for democratic governance in September 2009 raised local and international concerns about governance in Guinea, particularly in its mining sector. Still, very few academic and policy questions have been raised about either the structural foundations of poor political governance in Guinea or the relationship between political governance patterns and mineral resource governance, especially mining. Hence, it has become very important to understand the political and governance factors that have enabled Guinea to function as a legitimate state, despite 58 years of poverty and poor governance and 50 years of dictatorial rule.

Political instability in Guinea, if exacerbated, will have an impact well beyond the African region, as it will affect countries relying on Guinea for mineral resources, particularly bauxite, which is the resource for aluminium metal – the second-most-used metal in the world, used across a range of industries. Large-scale conflict in Guinea has the potential to affect the world market of bauxite since Guinea is amongst the leading producers and has the world's largest, highest quality reserves.

1.1.5   Central argument

The book's central argument is that the Touré (1958–1984) and Conté (1984–2008) regimes avoided large-scale armed conflict by using the bauxite industry as part of a broader coercive apparatus to strengthen themselves; this, in turn, facilitated the emergence of different forms of social contracts.

Under the first regime (1958–1984), led by Touré, mining enabled Guinea to form a key political alliance with the Soviet Union, which became Guinea's main political supporter and business partner. Russia's alliance enabled Touré to put in practice his nationalistic, dictatorial and anti-imperialist views, which helped the emergence of a 'strong coercive social contract' between the state and its citizens (discussed further in Chapter 3).

Under the second regime (1984–2008), led by Conté, the Guinean mining sector was liberalised, but this resulted in little socio-economic improvement amongst the general population. Conté was thus able to use revenue from bauxite mining to maintain regime stability by strengthening the military. This enabled Conté to overcome the threat to his regime when rebels from Liberia and Sierra Leone attacked Guinea, arguably avoiding a slide into civil war. Thus, during the Conté regime, a 'limited coercive social contract' emerged (discussed further in Chapter 4).

Industrial mining of bauxite contributed to regime stability through the provision of stable revenues to the two regimes. These revenues, along with the employment opportunities arising from the industrial mining of bauxite, enabled both regimes to pursue their ideologies and maintain their dictatorial nature. These events underscore the double-faced potential of mineral resources: capable of both triggering armed conflict in some circumstances and insulating countries against it in others.

Both the Touré and the Conté regimes neglected rural communities living in bauxite mining areas. The bauxite mining sector generates revenues for the state without the state having to make its presence felt locally. Consequently, in the case of CBG (Compagnie des Bauxites de Guinée), its direct contributions to local communities in the mining areas decreased pressure on the state to provide social goods to the communities in that region, thus rendering the state unaccountable to those communities. As a result, over the years, this has contributed to the emergence of an informal 'acquiescent permissive social contract', delivered by CBG as an intermediary between the state and the communities living in the regions where bauxite mining takes place.

The nature of these social contracts often reduced the potential for large-scale conflict. The 'coercive social contract' enabled regimes to use coercion to maintain regime stability, while the 'permissive social contract' (Diallo, 2017) emerged as a solution to restore order and maintain the status quo. As a result, a strong coercive social contract emerged under Touré, while a limited coercive social contract emerged under Conté. To restore order following periods of instability, what emerged in mining regions were an acquiescent permissive social contract and a 'circumstantial permissive social contract'. The acquiescent permissive social contract emerged where there was an informal ability for communities to bring and get social goods in returns for stability. The circumstantial permissive social contract emerged where the state finally let artisanal gold and diamond mining activities continue because it could not stop or control it and were unable to provide alternative opportunities for communities in resource-rich areas.

1.1.6 The paradox of plenty – the dream of freedom, colonial inheritance and continuing oppression, and exploitation and state dominance in Guinea

Allen (1995) suggests that it is important to avoid making general assumptions about politics and political systems in postcolonial African states because each state is different and has undergone various changes over time. The approach to understanding African politics should be grounded in an understanding and analysis of the historical context of specific countries, their colonial inheritance and the challenges they faced at independence.

Colonisation in Guinea started when the French armed forces defeated the forces of one of Guinea's last emperors – Almany Samory Touré – in 1898 (Pauthier, 2013). Although France faced internal resistance from the local Guinean population, by the early 1900s France had managed to gain full control of Guinea and its people (Gellar et al., 1994). As a French colony, Guinea joined French West Africa, known as l'Afrique Occidentale Française (AOF). Guineans became French subjects but with few rights: they were subject to forced labour and imprisonment without trial, and they had no right to popular representation or to vote (1994). During colonisation, Guinea was ruled by the French colonial administration, and France took control of the Guinean economy, its people and its natural resources. French colonisation meant that there was no emphasis on either building local capacity or promoting economic emancipation of Guineans. Therefore, when Guinea gained independence, it was a free country with no highly skilled labour.

Between 1945 and 1953, France implemented colonial reforms, which led to all Guineans becoming citizens of France, rather than its subjects, and Guineans were granted certain civil and political rights, including the right to form economic and cultural associations to promote their "interests" (Gellar et al., 1994:18). The newly gained liberty of Guineans, coupled with a boom in the mining sector, enabled the emergence of trade unions and other associations and created space for Guineans to voice their concerns about colonial rule (Okoth, 2006; Gellar et al., 1994). Between 1954 and 1957 the PDG (Parti Démocratique de Guinée), under the leadership of Sékou Touré, a popular trade unionist, emerged as one of the largest Guinean national movements with membership rising from "300,000 to 800,000" by 1957 (Okoth, 2006:42). The popularity of PDG was boosted by its anti-colonial stance. When Guinean nationals were granted the right to vote in 1957, the PDG won the majority of seats at the territorial elections. By 1958, the repressive nature of French rule over the years had resulted in strong anti-French movements, which mobilised people against colonial rule and promoted the fight for independence.

September 1958 is widely remembered as amongst the most important periods in Guinean history and politics. Guinea chose

> to opt for immediate independence by voting "no" in the September 28, 1958 referendum organized by General DeGaulle. Guinea was the only Black African Francophone territory to take this bold step despite warnings from DeGaulle of the adverse consequences that would follow from such a decision.
>
> (Gellar et al., 1994:19)

Adverse consequences included France's withdrawal from socio-economic activities in Guinea and the cessation of any financial aid to the country. In answer to France's warning, Touré became famous for having responded fiercely that, "We prefer poverty in liberty than riches in slavery".

By leading the opposition to French colonial rule, Touré was admired by many both in Guinea and across the African continent. However, Guinea's choice for independence from France had some negative consequences for the development of the Guinean state. Power under colonial rule had been centralised, and the French had made no effort to build the capacity of Guineans to manage their own state institutions. At independence, when France transferred political leadership of the country to Touré and cut its socio-economic support, a major poower vacuum was left to be filled by Guineans who had not been prepared or equipped to take over the skilled, technical and administrative senior and leadership roles vacated by French nationals. Touré inherited a country lacking skilled labourers, and there were very few economic activities that could function effectively. Although at independence Guinea had a vibrant agricultural and mining sector, it soon became reliant on just its booming mining sector, which enjoyed high international demand at the time (Okoth, 2006).

The period leading to independence and the consequent challenges presented to Touré shaped the nature of the newly independent Guinean state. This period heavily influenced the nature of international relations, the focus of the economy and the foundation of state authority, which emerged in the new Guinean postcolonial state. To secure state stability, Touré had to find external partners who shared his ideology and were ready to help him in his state-building efforts. Touré's ideology was based on "socialist principles which eventually evolved into a repressive and arbitrary personal dictatorship" (Gellar et al., 1994:20). In keeping with his ideology, at independence, Touré built commercial, social and economic alliances with Eastern Bloc countries (Lewin, 2010). This strategy had significant consequences for the development of the Guinean state.

Touré was left with no choice but to focus on the most profitable part of the economy, upon which he could develop commercial, technical and economic partnerships and which would generate revenues quickly to pay for the costs associated with state-building. Mining revenues offered 'windfall' gains, generated without the state having to work harder, invest, or develop any particular expertise – the mining companies paid taxes to the state and sourced and trained people to meet their own needs.

When Touré died in 1984, the military led a coup and took over the presidency. The coup leaders were organised under the Committee of National Recovery (CMRN). Lansana Conté became the head of state. The PDG and its constitution were dissolved. Despite making multi-party elections legal in 1993, Conté had full control over politics in Guinea. With support from the armed forces, Conté was able to maintain his hold on power, winning all subsequent elections until his death in December 2008.

When Conté first came to power, the CMRN promised socio-economic reforms that would improve the lives of Guineans; however, this never significantly materialised during his leadership. Although Conté liberalised the socio-economic spheres, towards the end of his reign, Guineans were still living in extreme poverty. From 2006 to 2008, there were ongoing protests demanding an improvement in the standard of living for ordinary citizens (Pauthier, 2013). By the time of Conté's death, despite 50 years of independence, no major improvements had occurred.

Both the Touré and the Conté regimes were repressive and dictatorial, yet both managed to remain in power until their leaders' deaths. The ability of both regimes to achieve this can be credited to their access to mining revenues, which provided the foundation for regime stability in Guinea. Karl (1997) argues,

> when minerals are the key sources of wealth for a state, these mining revenues alter the framework for decision-making. They affect not only the actual policy environment of officials but also other basic aspects of the state such as autonomy of goal formation, the types of public institutions adopted, the prospects of building other extractive capabilities, and the locus of authority.

(44)

Consequently, the dependence on mineral resources can weaken the state's capacities or strengthen them, depending on the vision and choices of its leaders.

Additionally, dependence on revenues from the extractive industry can often lead to three key challenges. First, states tend to focus on rent-seeking rather than developing other sectors, such as agriculture. Second, the collection of local taxes is neglected, making it difficult for the state and its civil services to be accountable for any shortcomings in socio-economic development. Third, capacity-building across state institutions is not addressed as a priority, because whether the institutions work well, or not, the state will still receive revenues. The risk is that without strong and adequate institutions, it is impossible to address shortcomings in socio-economic development within a country.

Because of Guinea's mineral endowment, one might have expected greater socio-economic development across the country. However, to Karl (1997), in the situation where the state depends on mineral extraction, "lessons from the past suggest a perverse relationship between some forms of natural-resource endowment and successful state-building" (242). It is this development failure that Karl (1997) calls "the paradox of plenty".

Unfortunately, in the case of Guinea, Touré inherited a nascent state with no strong local institutions; once in power, he focused on the security of the state and the stability of his political regime. In turn, Conté inherited a state with weak institutions; he focused on building the capacity of the armed forces and maintaining his hold on power. While Condé also inherited a weak state, instead of focusing on building strong institutional capacities to improve the governance of the mining sector, he focused his efforts on attracting new foreign investors in the bauxite mining sector. While this guarantees revenues for the state, it has undermined the focus on development of other sectors including agriculture and infrastructure (such as roads, water, electricity).

Additionally, in the absence of strong institutions, revenues generated from mining cannot effectively contribute to socio-economic development in Guinea. Although the current regime has implemented different initiatives to improve local development, local communities in bauxite mining areas are still living in poverty, as the whole country is. Today, despite promises to develop infrastructure, electricity, water and adequate health services are still inaccessible to the majority of Guineans. There are many ironies about Guinea: while it is recognised as the water reservoir of West Africa, with an abundance of natural resources, it remains one of the few countries in West Africa where there is no running water and electricity in the capital.

As Touré, Conté and the current president, Condé, discovered, the mining sector will generate revenues for the state without the state needing to intervene. The population is still hoping that these mining revenues will be utilised to transform the country. If we accept Karl's (1997) view, it is not impossible to reverse the perverse negative impact of having natural resources – the 'paradox of plenty'. In the case of Guinea, strong institutions can lead to improved socio-economic development in the future.

The book's arguments are discussed as follows.

Chapter 1 is the introduction, which has offered a general overview of the aims, objective, arguments, significance of the book, overview of Guinea's colonial inheritance and a note on the methodology.

Chapter 2 discusses the theoretical and conceptual framework of the study in greater detail. The chapter includes a review of the different literature and concepts related to the nature of postcolonial states in Africa, the resource curse debate, the concept of the successful failed state, the social contract in Africa and the 'rentier state' arguments. The chapter also offers an overview of the governance concept I use and the arguments linked to natural resource governance and its impact in Africa. The role and the impact of transnational organisations in the governance of the extractive sector in Africa are also discussed. The chapter ends with a conclusion on the contribution to the current literature to be made by this study.

Chapter 3 offers a detailed analysis of the nature of the Touré regime (1958–1984). The chapter explains how mining was used by Touré to maintain not only his political agenda and ideology but also the coexistence of regime stability and social insecurity. It explains how mineral resources enabled Touré to build political and economic alliances that helped Guinea to gain state legitimacy. This information will contribute to the understanding of how mining emerged as a tool that was used as leverage for dictatorship and authoritarianism in Guinea's politics. It also explains the emergence of the strong coercive social contract, which Touré maintained. This chapter uses data from archival, secondary and primary sources.

Chapter 4 offers a detailed analysis of the nature of the Conté regime (1984–2008). As in Chapter 3, this chapter sheds light on the role of mining in regime stability and social insecurity. The key challenges and impact of governance policies and reforms undertaken in Guinea's mining sector between 1984 and 2008 are assessed. It explains how mineral resources enabled Conté to fight major threats to his regime. The chapter also illustrates how frustration from unmet expectations for what the mining sector would provide to the population contributed to political tensions and threatened regime stability. The chapter also explains the emergence of the limited coercive social contract, which Conté maintained. This chapter uses data from secondary and primary sources.

Chapter 5 concentrates attention on the Guinean bauxite mining industry in detail. This section offers specific explanations about the contribution of bauxite mining to the socio-political and economic development of Guinea. The chapter focuses on a case study of CBG (Compagnie des Bauxites de Guinée). This chapter also offers insight on the emergence of an acquiescent permissive social contract that was facilitated by CBG despite the state's absence within the bauxite mining region. This chapter also discusses the impact of bauxite mining enclaves in Guinea.

Chapter 6 summarises the key research findings from the preceding chapters and highlights the key contribution and insights offered by the book and its research questions. It throws further light on how the mining of specific minerals, particularly bauxite, can contribute to regime stability and social insecurity. Finally, it also highlights the impacts of the emergence of different social contracts. It concludes with reflection on new efforts to improve the African mining sector since 2008 – not only in Guinea but also in the African Union, including the development of an African Mining Vision (AMV) as a mechanism for states in Africa to benefit more from the extractive sector.

1.1.7   Note on methodology

Fieldwork for the study includes interviews that took place in 2013, 2014, 2018 and 2019 in the bauxite mining regions and towns, including Boké, Kamsar and Sangaredi and the capital, Conakry. The interviews included individual interviews, group interviews, observations and focus group discussions with youth groups in the urban centre of Boké and the rural centre of Kamsar. The fieldwork included visits to bauxite mining companies, including Kamsar (2013, 2014, 2018 and 2019), Sangaredi (2013) and Boké (2018 and 2019). During these visits, staff from the mining company and local community members were interviewed to gain their view on the ongoing challenges and issues.

Research in the mining sector is complex. Before undertaking field research related to mining in Guinea, a researcher needs to have an official 'research permit' signed by the Ministry of Mines and Geology. Second, before undertaking interviews in the private mining companies, one needs to secure a letter from the managing director, which grants the researcher the right to undertake research and interview staff and/or access sites.

For this research, both secondary and primary data have been used. Primary data was collected during the field study through semistructured interviews, direct observation and analysis of primary documents from archives, private organisations, government institutions, NGOs and private businesses. Interviews were conducted with relevant stakeholders in rural, urban, mining and non-mining regions. An extensive review of secondary literature was undertaken using academic sources; policy documents; company, state and institutional reports; and online sources.

As a Guinean national, I speak two Guinean languages fluently in addition to French, and I understand the local dynamics, all of which make it easier to speak to a wide range of actors and travel between mining sites with limited risk. The field research was conducted in French, Fulani and Susu. French is the national language of Guinea, but in rural areas very few people speak French – instead, they speak Fulani and Susu. With the ability to speak local languages, I found that it was easier to build trust and that people were more likely to be open. Speaking the local language made the interviews seem less formal and intimidating, especially in rural settings. People felt they could speak more freely, and the risk of misinterpretation was reduced.

It was difficult to access data on the armed forces under the Conté regime, especially the military. There were two main factors contributing to this difficulty. The first is that, in Guinea, military information is considered sensitive and strategic and is therefore not publicly available. The second is that, from 1958 to 2008, the Guinean military exercised coercive power on the population, and major abuses of human rights perpetrated by them were recorded between the periods of 2006 and 2008. During this period, documentation about the military's activities was either not recorded or might subsequently have been destroyed. Thus, access to reliable data on the military for the period covered by this study has proven difficult; hence, I have relied on data from secondary sources.

One additional challenge that was difficult to overcome was trying to capture the views and experiences of women. Most interviewees were male. Since the research does not have a gender focus, the study reflects the reality on the ground – that the mining sector remains a highly male-dominated sector.

1.1.7.1   *A note on personal objectivity*

The qualitative nature of the research means that, as a researcher, I played a key role in the data collection and analysis process. Bourke (2014:2) suggests, "It is expected that the researcher's beliefs, cultural background (gender, race, socio-economic status, educational background) are important variables that might affect the research process". In carrying out my fieldwork and interviews, it was crucial for me to assess my own positionality as a multilingual female Guinean and as a UK PhD researcher/academic undertaking fieldwork in urban and rural areas across Guinea.

I am an urban, well-educated young woman from Guinea now living in the UK and working as an academic. The fact that I am from Guinea makes me an 'insider' to some extent – someone who shares the same nationality, religion and language as those of the majority of my respondents. On the other hand, during my research, I was an 'outsider' – someone who studies and currently lives abroad (Sanghera & Thapar-Bjorkert, 2008). This position comes with its own bias, which I have tried to minimise.

Positionality "represents a space in which objectivism and subjectivism meet" (Bourke, 2014:3). In addition, "field work is intensely personal, in that the positionality plays a central role in the research process in the field as well as in the final text" (England, 1994:251–252). Being aware of my positionality and the challenges and advantages that it presented allowed me to use my insider knowledge to access the research environment, while being as objective as possible as an outsider in providing accurate and unbiased research output. I was also aware that my background could affect my interaction with interviewees and their views towards me during the data collection in the field.

Being an insider has both advantages and disadvantages (Dwyer & Buckle, 2009). As an insider, one can have better access to local communities and be more accepted because of cultural understanding and communication facilities. Being able to speak with rural communities in their language makes it easier for research participants to share their experiences. As an insider, research participants sometimes identified with me. This can also have negative consequences for the research if the participants make incorrect assumptions that the researcher understands how they live and what they feel.

Doing research in the mining industry raises doubt and questions. My insider position was reinforced when, in some cases, I realised that some of the staff in office buildings I visited for the research were people whom I knew from my city in Kindia (Guinea); this was the case at the Ministry of Mines and Geology, for example. This helped participants feel at ease and more confident in the research intent.

My positions as both an insider and an outsider have informed the analysis and interpretation of my data. Without my ability to understand the language and other aspects of the cultural context in Guinea, I would probably not have been able to access or gain the co-operation of respondents. As a Guinean, I was able to easily interpret and understand what my participants were telling me. On the other hand, my university studies outside Guinea also enabled me to follow the research process and gather all the data needed to answer the research questions adequately and with more objectivity. In the end, I realised that although some aspects of my positionality, such as my gender, are fixed, other aspects, such as being an insider or outsider, varied, depending on the context of my fieldwork.

My gender and age affected the research process with certain actors in rural areas. I was a young woman undertaking interviews with in a male-dominated industry. In conversations with local communities, they saw me as a young Guinean woman who, despite all the challenges, had made an effort to visit these mining areas and had an interest in their livelihoods. As a result, when I started discussions, people felt more compelled to answer my questions and to 'help me'. At the end, many respondents concluded that it was very brave for a young woman like me to undertake such research, and the conversations in rural areas always ended with more encouragement and many blessings.

My gender and age also generated interest in me from the interviewees; hence, I often found myself having not only to ask questions but also to answer them. Respondents were often fascinated by my interest in mining and my 'courage', as most of them put it. This is due to the fact that most of the respondents had never come across a Guinean woman from a national or foreign university pursuing a PhD on mining or a young Guinean woman who was an academic in a foreign university. As a result, respondents often asked me many questions in return, questions not only about my research but also about my family, my childhood and my education in Guinea and abroad. These questions also became part of the field research process, and I often had to allocate additional time for these discussions.

No matter how stringent we try to be, absolute objectivity is a myth – there is always some background understanding relating to any empirical observation. It is therefore equally important to be aware of our subjectivity. To minimise subjective assumptions, I have from the outset undertaken this study with an open mind, a commitment to pay attention to detail and a determination to avoid drawing any conclusions not justified by the data. I sought clarification where things were not clear, and I tried to make sense of their responses from the perspective of their own subjective context as well as from the much wider perspective of the entire study context.

Note

1 Until 2003, the French company Pechiney owned this 10% of Halco.

References

Ake, C., 1975. A Definition of Political Stability. *Comparative Politics*. 7 (2), pp. 271–283.

Allen, C., 1995. Understanding African politics. *Review of African Political Economy*, 22 (65), pp. 301–320.

Andersen, D., Jørgen, Møller, Lasse, L.R., and Svend-Erik, S., 2014. State Capacity and Political Regime Stability. *Democratization.* 21(7), pp. 1305–1325.

Bannon, I. and Coller, P., 2003. *Natural Resources and Conflict: What We Can Do.* In: *Natural Resources and Violent Conflict: Options and Actions.* Washington, DC: World Bank.

Banque Centrale de la République de Guinée (BCRG), 2015. *Statistique, Tableau des Opérations Financières de l'Etat (1974–2000).* Conakry: BCRG.

Bourke, B., 2014. Positionality: Reflecting on the research process. *The Qualitative Report.* 19(18), pp. 1–9.

Campbell, B., 1983. *Les Enjeux de la Bauxite: la Guinée Face aux Multinationales de l'Aluminium.* Montréal: Les Presses de l'Université de Montréal et Genève.

Campbell, B., 1995. La Banque mondiale et le FMI: Entre la Stabilisation Financière et l'Appui au Développement. *Interventions Economiques* (26), pp. 111–140.

Campbell, B. (ed.), 2009. *Mining in Africa: Regulation and Development.* Ottawa: International Development Research Council (IDRC).

Campbell, B., 2010. Bonne Gouvernance, Réformes Institutionnelles et Stratégies de Réduction de la Pauvreté: sur quel Agenda de Développement? In: Jacques Fisette and Marc Raffinot, eds. *Gouvernance et appropriation locale, Au-delà des Modèles Importés.* Ottawa: Presse de l'Université d'Ottawa, pp. 35–56.

Chirico, P.G., Malpeli, K.C., Van Bockstael, M., Diaby, M., Cissé, K., Diallo, T.A. and Sano, M., 2014. *Alluvial Diamond Resource Potential and Production Capacity Assessment of Guinea* (ver. 1.1, April 2014): U.S. Geological Survey Scientific Investigations Report 2012–5256, p. 49. Available from: http://pubs.usgs.gov/sir/2012/5256. (Supersedes ver. 1.0 released 26 November, 2012.) [Accessed on 29 December 2018].

Collier, P., 2000. Doing Well Out of War: An Economic Perspective. In: Berdal, M. and Malone, D., eds. *Greed and Grievance: Economic Agendas in Civil Wars.* Boulder: Lynne Rienner.

Diallo, P., 2017. Social insecurity, stability and the politics in West Africa: A case study of artisanal and small-scale diamond mining in Guinea, 1958–2008,

The Extractive Industries and Society. 4(3), pp. 489–496, DOI:10.1016/j. exis.2017.04.003.

Dwyer, S.C. and Buckle, J.L., 2009. The space between: On being an insider-outsider in qualitative research. *International Journal of Qualitative Methods.* 8(1), pp. 54–63.

England, K., 1994. Getting personal: Reflexivity, positionality, and feminist research, *Professional Geographer.* 46 (1), pp. 80–89.

Gellar, S., Groelsema, B., Kante, M. and Reinstsma, M., 1994. *Democratic Governance in Guinea: An Assessment.* USAID Consultancy. Associates for Rural Development, 19 December 1994.

Huijbregts, C. and Palut, J.P., 2005. *Evaluation du Projet FSP: Decentralisation et Consolidation du Secteur Minier en Guinée.* Conakry Service de Coopération D'Action Culturelle à Conakry.

Hurwitz, L., 1973. Contemporary Approaches to Political Stability. *Comparative Politics.* 5(3), pp. 449–463.

International Crisis Group (ICG), 2007. *Guinée: Le Changement ou Le Chaos.* [Online] ICG. Available from: www.crisisgroup.org/~/media/Files/africa/westafrica/guinea/ French%20Translations/Guinea%20Change%20or%20Chaos%20French.pdf [Accessed on 22 May 2013].

International Monetary Fund (IMF), 2004. *Statement; and Public Information Notice on the Executive Board Discussion.* IMF Country Report No. 04/392. Washington DC: IMF.

International Monetary Fund (IMF), 2008. *Guinea: Poverty Reduction Strategy Paper, January 11, 2008, Poverty Reduction Strategy Papers (PRSP).* IMF Country Report No08/7. Washington, D.C: IMF.

Karl, T.L., 1997. *The Paradox of Plenty: Oil Booms and Petro-States.* Berkeley: University of California Press.

Knierzinger, J., 2018. *Bauxite Mining in Africa: Transnational Corporate Governance and Development.* Cham: Springer International Publishing.

KPMG, 2014. *Guinea Mining Guide.* KPMG International.

Labbé, F.J., 2004. *Objet: Les Revenus Minier de la Guineé Baissent-ils? Lesquels et Pourquoi-Element de Reflexions.* Conakry: Ministère des Mines et de la Geologie (MMG).

Lewin, A., 2010. *Ahmed Sékou Touré (1922–1984): Président de la Guinée de 1958 à 1984.* Paris: L'Harmattan.

Ministère de l'Economie et des Finances (MEF), 2002. *Stratégie de Réduction de la Pauvrété en Guineé.* République de Guineé, Conakry: Ministère de l'Economie et des Finances.

Ministry of Mines and Geology (MMG), 2005. *Guinea, Mineral Resources, Bauxite.* Republic of Guinea, Conakry: MMG.

Ministry of Mines and Geology (MMG), 2018. *Invest in Guinea.* Sahel and West Africa Mining Conference, London. 8–9 May 2018.

Okoth, A., 2006. VI, *A History of Africa.* Kampala, Dar es salaam: East African educational Publishers Nairobi.

Oliveira, R.M.S.D., 2007. *Oil and Politics in the Gulf of Guinea.* Columbia/Hurst: Columbia University Press.

Pauthier, C., 2013. L'héritage controversé de Sékou Touré, "héros" de l'indépendance. *Vingtième Siècle: Revue d'histoire.* 118(2), pp. 31–44.

Programme d'appui à la gouvernance dans le secteur minier (PAGSEM) by EGIS International, 2016. *Etude Stratégique Environnementale et Sociale (ESES) de la réforme du secteur minier en République de Guinée,* Rapport FINAL Définitif, 8 April 2016. P. 13.

Ross, M.L., 1999. The Political Economy of the Resource Curse. *World Politics.* 51(2), pp. 297–322.

Sanghera, G. S., and Thapar-Bjorkert, S., 2008. Methodological dilemmas: Gatekeepers and positionality in Bradford. *Ethnic and Racial Studies.* 31(3), pp. 543–562.

Touré, K., 2013. *Compagnie des Bauxites de Guinée (CBG), 2013 – A Guinean Success Story,* June 2013. London: CBG.

U.S. Geological Survey (USGS), 2011. 2009 Mineral Year Book: The Mineral Industry of Guinea. Bermúdez-Lugo, Omayra: US Geological Survey Mineral Year Book-2009. Available from https://s3-us-west-2.amazonaws.com/prd-wret/assets/palladium/production/mineral-pubs/country/2009/myb3-2009-gv.pdf

U.S. Geological Survey (USGS), 2018. M*ineral Commodity Summaries: Bauxite and Alumina,* January 2018. Available from: https://s3-us-west-2.amazonaws.com/prd-wret/assets/palladium/production/mineral-pubs/bauxite/mcs-2018-bauxi.pdf [Accessed on 29 June 2019].

United Nations Development Program (UNDP), 2018. *Human Development Indices and Indicators: 2018 Statistical Update; Briefing Note for Countries on the 2018 Statistical Update; Guinea.* Available from: http://hdr.undp.org/sites/all/themes/hdr_theme/country-notes/GIN.pdf [Accessed on 17 April 2019].

2 Framing Guinea
Mineral extraction, rent and stability

2.1 Introduction

This chapter reviews and introduces a theoretical framework for explaining how resource governance has contributed to reshaping the political dynamics of state-society relations in Guinea. The chapter commences by discussing the diverse nature of postcolonial states in Africa and then proceeds to identify the types of political regimes widely perceived to have emerged during this period. It introduces the concepts of concurrent stability and insecurity within a state, as referred to in this study. Various studies attempting to address the challenges to project state authority in certain African countries are discussed, and the concept of social contract, as it relates to Africa in general and Guinea in particular, is introduced.

The resource curse concept is introduced in this chapter to enable an understanding of socio-political theories linking natural resources and conflict in Africa. The linkages between the extractive sector and conflict in Africa are explored by looking at the case of Sierra Leone, one of Guinea's neighbours. While the conflict in Sierra Leone was linked to mineral resource, Guinea managed to avoid such a conflict despite enormous mineral wealth. It is expected that this examination of the Guinea case will offer an interesting perspective on how countries might be endowed with similar types of mineral resources, fraught with the same socio-political dynamics, but necessarily produce different outcomes, particularly concerning regime stability.

2.2 Postcolonial states in Africa: Guinea and the discourse of neopatrimonialism and state collapse

The postcolonial state in Africa is a recurrent topic in African studies (Adedeji, 1993; Chazan et al., 1992; Herbst, 2000; Nugent, 2004). Most authors suggest that the African state is poorly emancipated from society and therefore patrimonial, although there is a lack of agreement on this

(Chabal & Daloz, 1999). Authors use neopatrimonialism to explain how African leaders maintained the coexistence of formal and informal politics within their states (Englebert & Dunn, 2014; Reno, 2000; Ahluwalia, 2001; Chabal & Daloz, 1999; Bayart, 1993). Neopatrimonialism has created a space "where formal constitutions and organizations are subordinate to individual rulers (the president or the 'big man') and personal relationships are the foundation and superstructure of political institutions" (Szeftel, 1998:235).

In the 1990s state collapse became a feature of many African states, including Liberia, Sierra Leone, Somalia, Chad and Zaire. These countries witnessed the emergence of 'warlord politics', where politics in Africa was seen as linked with crimes in those countries (Cliffe & Luckham, 1999; Goulding, 1999 in Ahluwalia, 2001:65–66; Reno, 1995). To Zartman (1995:5) a collapsed state is characterised by the inability of the state to perform three crucial functions: implementing laws and maintaining order, maintaining legitimacy and managing public affairs. The objective of the warlord leader is to stay in power and build up wealth and opportunities that will sustain his network and supporters.

Most authors view neopatrimonialism as a form of governance that had stifled the growth of effective state regulatory institutions. What is seen externally to make the state in Africa weak is also what maintains it within African societies (Chabal & Daloz, 1999). Many rulers have shown no interest in strengthening institutions within their state (Reno,). Weak formal institutions benefit African rulers and their supporters because they enable them to maintain their grip on power. Since independence, neopatrimonialism has been a key feature of regimes in Guinea. To date all regimes have been concerned in maintaining power. Both Conté and Touré died in office, and Condé is currently attempting to modify the constitution to enable him to stay in power for three terms, although the majority of the population does not support this.

2.1.1   *Regimes in post-independence Guinea*

The apparatus of the state manifests itself through different political regimes and governments. Regimes are "concerned with the form of rule and interaction between citizens and the state agencies. If the idea of the state is associated primarily with organization of power, regime focuses on how state power is exercised and legitimated" (Chazan et al., 1992:39). Government relates to

> Institutions responsible for making collective decisions for society. . . .
> In popular use it refers just to the highest level of political appointment:
> to presidents, prime ministers and others at the apex of power. But in a

wider conception, government consists of all the organizations charged with reaching and executing decisions for the whole community including the police, the armed forces, public servants and judges. . . . It is the entire terrain of institution endowed with public authority.

(Hague & Harrop, 2013:4)

The following section offers an overview of state and regime typologies that have emerged in post-independence Africa. It also identifies the type of regimes encountered in Guinea (1958–1984). Chazan *et al.* (1992) identify six major types of regimes in Africa excluding South Africa:

> Administrative-hegemonic regimes: the main policy decisions are centralized around the leader and his close advisors; Pluralist regimes: activities have been a mixture of bargaining, compromise, and reciprocity; Party-mobilizing regimes include Guinea (1958–1984) and are discussed in subsequent paragraphs. Regimes in this category, have a strong socialist predispositions; Party-centralist regimes: exercised with absolute central control and direction with less tolerance of accommodation with local social forces or with most external actors; Personal coercive regimes: the entrenchment of the regime has been predicated on the connection between a strong leader and the coercive apparatus as was the case of Guinea under Touré; Populist regimes: a concept of social inclusion defined in non-elite terms.
>
> (pp. 136–149)

This typology of regime types provides a useful structure to introduce and analyse Guinea's political history. In Guinea, the Touré regime (1958–1984) varied over time. After Guinea's independence in 1958, political parties in Guinea were merged with the PDG (Parti Démocratique de Guinée), which became a *de facto* single-party state in Guinea (Nugent, 2004:193). The PDG was a combination of "one-party state with socialist doctrines" (Ahluwalia, 2001:66). The Touré regime in Guinea from 1958 to 1984 was mostly in line with the description of the party-mobilising regime:

> Regimes in this category reflect . . . strong socialist predispositions. . . . Public institutions in these regimes are rested on a combination of a strong one-party domination coupled with bureaucratic expansion under the control of an executive president. . . . Thus the politburo of Guinea's Party Démocratique de Guinée (PDG) . . . which retained a decisive capacity to shape public policies, discipline party members, and process appointments to high executive and civil service positions. . . . In these cases, coercive devices have been used to consolidate party-state control.
>
> (Chazan et al., 1992:142)

In its later years, the Touré regime turned into a personal coercive regime, which is "the most unpredictable type of regimes because the political processes and procedures can change depending on the mood of the leader who in this case has power over all structures of the state" (Chazan et al., 1992:148). The PDG claimed to promote national unity and development. Despite these claims, the Touré regime was coercive and took no account of citizens' views. It was rather divisive in that it eroded trust amongst citizens, and people were scared to criticise the party, as they could risk a prison sentence or execution (Rivière, 1977). Citizens had no voices of their own; any voice represented merely that of the party in power, and anything spoken against the party risked repressive actions. This led to widespread accusations and frequent brutalities across the country, led by the PDG.

Touré's death in 1984 led to a power vacuum, where the PDG could not agree on a successor (Nugent, 2004:195). As a result, a new military junta took power, and Colonel Lansana Conté became the new head of state, remaining in power from 1984 to his death in 2008. After Touré's death, part of his legacy remained with Conté, whose main goal was to remain in power. Conté continued to use coercive measures, although to a lesser extent than Touré, to hold on to power. The Conté regime was not a party-mobilising regime – the regime presented forms of a personal coercive regime. Conté later became a civilian ruler, but he continued to lead the country with a high reliance on the military apparatus, which was used to safeguard national security and oppress citizens.

As opposed to Touré, Conté embraced some of the neoliberal ideologies of Western donors (cf. Chapter 4), but the nature of his regime remained dictatorial, and he made sure that no structures threatened his position as president. Although the Conté regime was less oppressive than that of Touré, "in Guinea a culture of authoritarianism remained deeply ingrained" (Nugent, 2004:396).

In 2010, after the period primarily discussed in this book (1958–2008), Guinea appeared to have its first democratically elected president, Alpha Condé. The Condé regime shares the features of the "administrative-hegemonic regimes: the main policy decisions are centralised around the leader and his close advisors" (Chazan et al., 1992:136–149). Since 2010, politics in Guinea have been linked with ethnic division and concerns over the unprecedented expansion of the mining sector, which on many occasions has fuelled claims of corruption and mismanagement.

Despite the coercive nature of the Touré and the Conté regimes, both were able to maintain the coexistence of regime stability and social insecurity (cf. Chapters 3 and 4). This can be credited to the fact that both retained control over the security forces and the mining sector. However, Touré had more control over the mining sector than Conté – while Conté adopted a liberalised approach, Touré closely controlled all economic sectors in Guinea.

The next section discusses the challenges faced by African regimes in their attempt to project their authority across states.

2.1.2 State, politics and projection of authority across African post-independence territories

To Chazan *et al.* (1992:133), "the dynamic of politics in Africa is about the procedures and mechanisms by which state agencies and social groups cooperate, conflict, intertwine and consequently act". Chazan, *et al.* (1992:159–220) and Le Vine (2004) identify three domains of politics within a state. The first domain of 'high' politics (politics of the state) happens at the macro level, within institutions at the national level. The second type of politics is the 'low' or deep politics, and this is politics of how citizens relate to the state and its institutions. The third type is the domain of 'para-politics' or nonformal domain, where the majority of daily political exchanges happen in Africa (Le Vine, 2004). The 'para-political' space is where the vast majority of informal business, illicit activities and personal politics takes place, and it is often beyond the control of state structures (Le Vine, 2004:305). The different domains of politics and political actors present a real challenge to African states when trying to project their authority beyond formal areas to para-political domains.

The interaction between low politics and para-politics domains are of particular interest here because the majority of mining activities take place in rural areas. However, most of the governance initiatives are decided from the capital, where high politics tends to take place. Hence, it is important to understand how African leaders attempt to project their authorities across given countries.

In an attempt to offer clarifications on the strategies used by African leaders to project their authority, Boone (2003) offers a good explanation of the relationship between African states and their rural populations. Through the analysis of six case studies – including Senegal's groundnut basin and the Casamance, Ghana's Asante region, the Senegalese River Valley and Côte d'Ivoire's Korhogo and southern regions – she tries to address the question, "how is politics configured at the local level? How do rulers choose strategies for governing the countryside, and when do strategies change" (2003:1). In explaining the unevenness in the involvement of rulers, her main argument is that the state engages in select areas where there are organised rural elites that force it to do so. As a result, the state provides rural areas with development of local infrastructures in return for access to resources or opportunities that exist in those rural areas (2003).

As opposed to most West African countries, chieftaincy was abolished in Guinea on the eve of independence; hence, there is very little hierarchy of power in rural communities, and there are no organised rural elites. Although there are respected religious elders, they serve more as preachers of peace and stability in the community rather than organised elite groups whose role is to bargain for improvement of local infrastructures. In Guinea, the absence of rural elites meant that the state could afford to ignore the rural areas because there were no organised structures that could bargain for development on their behalf. This is one reason why resource-rich areas can remain amongst the poorest regions in a country. It is exactly the case in diamond mining areas in Guinea, where despite being rich in resources, they are often neglected by both the state and the diamond mining companies, with no infrastructure such as hospitals, clean water and/or electricity (Diallo, 2017).

Herbst (2000) tries to explain the challenges faced by African leaders within their attempt to control and consolidate power across given state territories. To Herbst (2000:2), "states are only viable if they can control the territory defined by their borders. Control is assured by developing an infrastructure to broadcast power and by gaining the loyalty of citizens". Unlike Herbst, Boone (2003) does not assume that African leaders intend to control their territories. She sees the lack of development of certain rural areas in Africa as a strategic choice by leaders rather than a weakness. What Herbst (2000) highlights is the danger of the inability of state to control its territories and the challenge faced by the state in trying to maintain populations across different landscapes throughout the country.

I align to some extent with Boone's (2003) suggestion that African leaders choose to leave certain rural areas to their own devices by choice. However, neither Herbst (2000) nor Boone (2003) mentions how regimes with strong military support and mineral resources project their authorities across African states (Nugent, 2004) or highlight how, sometimes, states build infrastructure to maximise their revenue from mining activities. By studying Guinea, this book will contribute to fill this gap. The next section will discuss the notion of *regime stability* and *social insecurity* and how they are used in the context of this book.

2.3   The vicissitude of regime stability and social insecurity in Guinea

Ake (1975) sees political stability as the absence of structural change within a given political system. According to Hurwitz (1973), there is little agreement on the concept of political stability because there is even less agreement on the term 'stability':

> The differing views and approaches to political stability are seen to be: (a) the absence of violence; (b) governmental longevity/duration;

(c) the existence of a legitimate constitutional regime; (d) the absence of structural change; and (e) a multifaceted societal attribute.

(449)

To Andersen *et al.* (2014),

> State capacity is what reinforces regime stability and state capacity can be perceived as the extent to which the state possesses the coercive capacities (also known as monopoly on violence) in autocracies and administrative capacities (also known as administrative effectiveness) to penetrate and regulate society and extract and appropriate resources in democracies.

(1305–1306)

For this book, the notion of regime stability draws from the work of Andersen *et al.* (2014), Hurwitz (1973) and Ake (1975). Regime stability focuses on a combination of different attributes. Regime stability is taken to mean a) governmental longevity/duration, b) the absence of structural change and c) the ability of the state to possess coercive capacities and use them to minimise any threat to its longevity. In the case of Guinea, the Touré and the Conté regimes lasted 26 and 24 years respectively. Regime stability does not necessarily translate into a positive impact on the wider society or the complete absence of threats, pockets of protests and violent confrontations. Regime stability in Guinea has been retained with different forms of oppression and restriction on civil liberties. While regimes focused on maintaining regime stability, the socio-economic development of the population was neglected, thus leading to social insecurity, discussed in the following sections.

Initially, the concept of insecurity was mainly addressed from scientific disciplines such as psychology, social psychology and psychiatry (Vornanen et al., 2012:281). Cameron & McCormick (1954) were the first authors who attempted to analyse the concept of insecurity from a social science perspective. They concluded that there were "no consistent definitions or theories" that existed for this concept and "challenged all genuinely interested social scientists" to undertake further field research that would help clarify the concept (1954:561). Several years later, Bar-Tal and Jacobson (1998:69) suggest that it is important to consider psychological variables when analysing insecurity, as it enables one to understand insecurity from the perspective of citizens within a society. They add that understanding feelings of insecurity needs to be put in context and time-bound (Bar-Tal & Jacobson, 1998). However, there is still no clearly defined and agreed-upon concept of social insecurity within the political and the social science literature. The concept of social insecurity does not seem to have been well-developed within the social sciences. In the case of Guinea, this book explains the concept of social insecurity in relation to the local context of the Guinean society and everyday interactions between rulers and citizens. Hence, it becomes a question of how the political field relates to the socio-economic field.

This book takes the concept of social insecurity as a notion that arises from a lack of socio-economic development, which, in turn, has resulted in low living standards and limited employment opportunities for the majority of Guineans. Social insecurity is taken to involve poor socio-economic development, including a) low living standards, b) increasing poverty rates, c) poor development of social infrastructure and d) unemployment. The unavailability or unequal/unreliable distribution of social goods has resulted in an increasing rate of corruption and poverty across the country, as seen under the Touré, the Conté and the Condé regimes; in extreme cases, this has led to social uprising and protests. It has also increased rates of poverty, which have affected Guinea throughout its post-independence history despite the country's wealth of mineral resources.

In Guinea, as advanced in this book, regime stability is not incompatible with social insecurity – in fact, both have emerged and functioned for over 50 years. The relationship between regime stability and social insecurity is neither linear nor causative; it is complex, and it can be co-constituted, paradoxical or even unrelated. The survival of regimes in Guinea can be credited to their ability to maintain the delicate coexistence of regime stability and social insecurity.

This book claims that social contracts are what enable the two to coexist. The social contracts regulate how the political field relates to the socio-economic field. In the case of Guinea, funtioning social contracts are facilitated by the country's mineral resource endowment. The next section discusses how mineral resource endowment has contributed to the ability of regimes to maintain state legitimacy and reinforce the relationship established between rulers and the ruled in Africa.

2.4   The social contract and societies in Africa

This section focuses specifically on explaining social contracts as developed by Nugent (2010). To try to understand the weaknesses and challenges of states that have emerged in the postcolonial era, several scholars have focused on how leaders use power, domination and neopatrimonialism to maintain state legitimacy. To Englebert and Dunn (2014:60), "the exchange of state-provided development for political allegiance can be thought of as the founding social contract of the post-colonial era".

According to Nugent (2012:1), "a social contract does not imply an inherently harmonious relationship, merely with those who exercise power and those who on the receiving end share a common understanding of what that relationship entails" (Diallo, 2017). Nugent (2010, 2012) identified four types of social contracts that legitimated the African state at independence: coercive, liberational, productive and permissive social contracts. The coercive social contract emerges and is sustained under regimes that often resort to the use of force, threats or promises of protection. In a 'productive social contract', there is a consensus between the ruler and the ruled, which results in the effective delivery of state services to the populations in exchange for their obedience to state rules and regulations (2010:43). In a permissive social contract, rulers allow communities to undertake activities that are not legal in return for accepting their rules and regulations, which "often involves negotiation over payment of taxes as well as rights of access to scarce resources, including land" (43). A 'liberational social contract' emerges as a result of national liberation processes or against extreme dictatorships (2010).

In Guinea, coercive and permissive social contracts explain the dynamics of state–society relations and the consolidation of political power by successive regimes. This book expands Nugent's (2010) typologies to account for a detailed explanation of such dynamics and suggests the following subcategories under coercive social contract and permissive social contract:

- *Strong coercive social contract:* This term is used to highlight regimes that have resorted to excessive coercion to govern society. In this case, the state uses coercive measures to control and discipline citizens. It leaves citizens no choice but to obey the authority of the regime to avoid facing repression from state forces. This situation will be used to describe the relationship between the state and its society under Touré and how mineral resources were used to reinforce this relationship (cf. Chapter 3).
- *Limited coercive social contract:* This refers to a situation where citizens have to obey the state but at the same time they have some flexibility in their interaction with the state and are free from cruel oppression. This situation is further developed in the book to describe the state-society relationships under Conté (cf. Chapter 4).
- *Circumstantial permissive social contract:* This portrays a situation in which, given the circumstances, nothing can be done to force the population to respect state legislation. In the context of mining activities, this is a scenario where nothing can be done to stop the mining activities; hence, they continue despite activities being legally banned or prohibited by the state. This can be used to describe the relationship between the state and artisanal diamond mining actors (Diallo, 2017).

- *Acquiescent permissive social contract:* This describes a situation where activities happen as agreed to by the state under terms and conditions consented to by both the state and its citizens. However, in some cases, this agreement is informal, and it is negotiated in return for stable mining activities. This explains the relationship between the state, local citizens and bauxite mining companies.

The way some of these contracts played out in Guinea's post-independence history will be discussed in different chapters of the book. The next section discusses the literature on the resource curse, argues for a tenuous link between mineral resource endowment and instability and emphasises the stabilising impact of resource abundance.

2.5   The resource curse and instability in Africa: comparing Guinea to Sierra Leone

The resource curse is explained as "the inability of resource-rich countries to unlock mineral wealth for the benefit of the citizenry of developing countries" (Auty, 1993, cited in Hilson & Maconachie, 2009:53). Resource curse arguments have been focused on the impact of resource extraction on political regimes, economic performance, civil war and the nature of governance in different countries. The different arguments on the resource curse can be classified into three main subgroups.

The first group suggests that there is a link between resource wealth and economic performance, and that those countries that are highly dependent on mineral resources are often affected by poverty and low economic growth compared to the average (Sachs & Warner, 1995, 1997, 2001; Ross, 1999). The second group argues that there is a link between natural resource dependence and the nature of political regimes and suggests that the emergence and the survival of authoritarian regimes, dictatorship and democratic breakdowns are predominant in countries rich in natural resources (Jensen & Wantchekon, 2004; Ross, 2001; Wantchekon, 2002; Reno, 2003). The third group argues that there is a link between dependency on natural resources, especially minerals and oil, and conflict causation in developing countries, as they tend to be affected by civil war or widespread poverty or both (Basedau & Richter, 2011; Collier & Hoeffler, 1998, 2004; Bannon & Coller, 2003; Collier, 2000; Keen, 1998).

Basedau and Richter (2011) offer a good summary of the three main causal mechanisms identified across the academic literature that underline the link between natural resources and violence leading to conflict (Humphreys, 2005; Ross, 2004; Le Billon, 2008, cited in Basedau it is Basedau and Richter, 2011, 2011:6). First, the motivation of insurgents to take up arms might be a consequence of resource-related environmental damage or situations in which people feel that they are not benefiting from the resources. Second, resources can facilitate conflict by providing a financial source to sustain armed conflict through the 'lootability' of resources by rebel groups. Finally, natural resource abundance might have hurt state institutions (weak state) and the economy (Dutch disease) given the tendency to depend on rents from these resources while neglecting the importance of building strong institutions capable of generating and distributing resources. As a result, reliance on weak institutions and poor economic growth increases the likelihood of social unrest.

Several authors suggest that the impact of contextual conditions is pivotal for the presence or absence of the resource curse and thus argue that these contexts need to be accounted for in studies linking mineral resources and conflict (Basedau & Lay, 2009; Snyder & Bhavnani, 2005; Basedau, 2005). This book is in line with the argument of these authors in that, where the resource curse is present, this depends on specific contextual realities. For instance, the presence of lootable resources in countries with low capacity to police their frontiers are prone to threats of instability (Snyder, 2006; Snyder & Bhavnani, 2005:6).

Between 1991 and 2002, Sierra Leone went through a brutal civil war. It started when the Revolutionary United Front (RUF) invaded the Eastern Region of Sierra Leone from Liberia in 1991 (acaps, 2014:2). Led by Foday Sankoh, a proxy of Charles Taylor (then Liberia's president), the RUF was composed of radical students, marginalised rural youth and informal diamond diggers who intended to overthrow the government (Keen, 2005; Hanlon, 2005). The war resulted in the about 70,000 deaths (Gberie, 2005). The RUF cut the limbs off an estimated 3,000–9,400 people (Guberek et al., 2006:26). The war displaced 1.5 million Sierra Leoneans, amongst whom 1 million went to Guinea (Guberek et al., 2006:24). The war destroyed most of the country's infrastructure and reversed the country's development. Various arguments were put forward to explain the causes of the conflict, but a detailed analysis of the causes of the civil war in Sierra Leone is beyond the scope of this chapter.

Certain authors believe that the civil war was a result of the collapse of the neopatrimonialist system that had been in place since independence (Reno, 1998; Richards, 1996). Others have linked the conflict to the 'youth crisis' in Sierra Leone, which has been linked to the state's failure to provide adequate socio-economic development and employment opportunities for the country's youth (Hoffman, 2006; Richards, 2006; Kandeh, 1999; Abdullah, 1998; Bangura, 1997; Rashid, 1997). These authors argue that it was easy to manipulate youths to join the rebellion as most of those in rural areas were unemployed, uneducated, marginalised and excluded from the government elites (Gberie, 2005). Collier and Hoeffler (2000, 2004) argue that 'greed', rather than 'grievance', is mainly what motivates rebels to go to war. They suggest that because rebellions need financial resources to sustain themselves, the potential profit of looting diamonds in a country like Sierra Leone was perceived as an attractive option for youths who had no education opportunities and were unemployed. These authors see economic incentive as the main reason for rebellions, as opposed to legitimate socio-political grievances.

In resource-rich countries, the failure of the state to meet the socio-economic needs of its citizens can lead to political and economic grievances that can increase the likelihood of violence (Ballentine & Nitzschke, 2005). Those who suggest that grievance is one of the main causes of civil war have taken some of these factors into account. As a result, other authors have argued that grievance or 'justice seeking' is what causes conflict. To Stewart (2008), the causes of civil war are found in a mix of horizontal inequalities, which include economic, political, social or cultural inequalities amongst groups. In the case of Sierra Leone, the collapse of the state, the availability of lootable diamonds, the youth crisis and existing grievances spurred the civil war and the capacity of rebel leadership to enlist youths in the rebellion.

In Sierra Leone, the geographic accessibility of diamonds, the absence of an effective state presence in diamond-rich rural areas and the lootability of diamonds by rebel groups facilitated the use of diamonds to finance and sustain the conflict (Basedau & Richter, 2011; Basedau, 2005). Lootable resources with low economic barriers can be seen as an easily accessible resource for rebels – if unprotected, ease of access to these areas can threaten political stability (Snyder & Bhavnani, 2005:6; Snyder, 2006). If it were kimberlite diamonds (as opposed to alluvial diamonds), which are 'non-lootable' because mechanised processes are needed to extract them, one could argue that it would have been more difficult for rebels to establish control over these diamond mining sites for a longer period or enlist as many youths as they did. Greed alone is not enough to hold a rebellion together for a decade or even start a conflict without grievance. Grievance alone is not necessarily a reason why conflict starts. Guinea is a case in point – it has both lootable alluvial diamonds and non-lootable bauxite resources, and despite the ongoing popular grievances as a result of corruption, increasing poverty, poor governance and nepotism, the country did not go into mineral resource–linked large-scale conflict between 1958 and 2008. This is despite the ongoing conflicts in Sierra Leone, Liberia and Ivory Coast.

Although rural areas in Guinea were as poor as those in Sierra Leone, the main difference between Guinea and Sierra Leone was state control and presence. As opposed to Sierra Leone, diamond mining areas in Guinea have always been subject to a heavy military presence loyal to the ruling regime. This would have made it difficult for any external forces to take over diamond mining sites without resistance. Guinea contains stores of both lootable (diamonds) and non-lootable (bauxite) resources. In conclusion, various mechanisms come into play before conflict manifests itself. These include both non-resource-related and resource-related variables. More importantly, conflict emerges where social contracts have broken down and rulers and those ruled no longer find a bargaining point that is considered beneficial to both parties.

The 'greed versus grievance' thesis is hardly a sufficient explanation of the causes of civil wars. By using bauxite mining in Guinea as a case study, this book hopes to illustrate the impact of non-lootable resources on regime stability and social insecurity. Bauxite – unlike gold or diamonds – is not intrinsically valuable in ore form. It requires energy-intensive transformation to create aluminium; thus, Guinea's bauxite is effectively 'theft-proof'. However, as discussed in Chapter 5, because of the logistics and infrastructure implemented for the extraction of bauxite, mining companies can incur major losses when disrupted by protests. As a result, bauxite mining companies often find quick solutions to local protests as soon as they start, thus preventing local grievances from turning into large-scale conflicts. The next section moves away from the impact of mineral resource wealth on conflict to its affect on stability.

2.6 Beyond the resource curse: from instability to stability

Not all countries rich in natural resources are affected by conflict, authoritarianism and/or slow economic growth. Norway, Chile and Botswana are "stable democracies that are economically prospering" (Mähler, 2009:8). Recently, groups of researchers have been trying to explain the stabilising impacts of resource wealth that have until recently been ignored by most authors (Dunning, 2008; Snyder & Bhavnani, 2005).

According to Snyder and Bhavnani (2005:4–5), mineral resources are likely to contribute to stability if they are non-lootable (e.g. bauxite, iron). They contribute to state revenues through taxes, and the state uses these revenues to strengthen security forces and contribute to the state's expenditure on social welfare. Thus, these limit grievances and the ease of recruitment of rebels. On the other hand, lootable resources are difficult to control because they can be easily extracted and thus easily controlled by rebels in the absence of state presence (2005). The authors believe that revenue is the source of stability of the state and that the chances of state collapse are higher when a state has no revenue – that is, "no revenue, no regime" (10).

Botswana is a good example of a mineral resource–rich African country that has not fallen victim to conflict, poor economic growth and corruption. It is cited as the main exception to the resource curse in Africa. It has managed to successfully exploit its diamonds. In Botswana, the progress of development is credited to the discovery of non-lootable kimberlite diamonds, which are deep underground and need major infrastructure for extraction. In addition, Botswana has had stable political leadership and policies, the presence of good institutions, governance, successful public and private partnerships, a favourable business environment, prevention of ethnic tensions, effective management of mining revenues and natural advantages (Van Wyk, 2009:17; Acemoglu, Johnson & Robinson, 2001).

Although other countries in Africa with non-lootable resources – such as Nigeria, Equatorial Guinea, Guinea and Niger – have similar advantages, poor governance, corruption and widespread poverty have affected them. Botswana's governance of the mining sector, supported by good institutions and the way revenue was spent, was crucial for the success of mining (Kowalke, 2009; Acemoglu, Johnson, & Robinson, 2001). With non-lootable resources that allow leaders to easily collect taxes, Botswana can choose to invest these funds either in the development of the public sector or within its patronage systems. In most countries, the latter happens. Botswana shows that, with good institutions, mineral resources can contribute to economic growth and development in Africa.

Since regimes need revenue to survive, revenue from mineral resource extraction is often used to maintain survival; regimes use this revenue to contribute to security expenditure and to welfare initiatives (Snyder & Bhavnani, 2005). The rentier states also show that countries have been able to use oil rent revenue to maintain regime stability. Yates (2009:7) defines a rentier state as "any country that receives regular substantial amounts of external rent". Rentier state "regimes use revenue from abundant resources to buy off peace through patronage, large-scale distributive policies and effective repression" (2009). As a result, these states appear to be "politically stable and less prone to conflict" (Basedau & Lay, 2009:2).

However, this rentier thesis does not give sufficient explanation as to how oil-rich countries manage to survive despite the ongoing internal challenges some of them face. Soares de Oliveira (2007) expands the rentier state argument to account for a crucial element: the continued international recognition of resource-rich states that otherwise fulfil all the categories that would usually cause states to be branded as 'failed' or 'fragile' and thereby in need of significant external intervention. Through case studies of countries like Angola and Nigeria, Soares De Oliveira (2007) investigates the link between oil and politics in the Gulf of Guinea and how this has formed states that are failing yet surviving.

For African countries like Nigeria, Equatorial Guinea, Angola and Chad, the revenues derived from the extraction of oil allow their governments to be financially independent from domestic taxpayers and thereby unaccountable to their citizenry (2007). Soares de Oliveira observes that oil "allows states to survive regardless of bad policies, permit elite material success regardless of reckless management, earn international allies regardless of domestic conduct, and make companies want to invest regardless of risk" (329). These successful failed states, he argues, have failed internally because socio-economic development is lacking: institutions to ensure the welfare of citizens hardly exist, a small elite group controls the oil revenue and government is not or cannot be held accountable. Yet these states are successful at surviving externally because their governments are skilfully using the legal sovereignty over the valuable commodities at their disposal to bargain for their continued recognition on and valuable access to the stage of international diplomacy and transnational business (2007).

Soares De Oliveira offers an excellent analysis of the positive impact of oil in the way that it enables the survival of oil rich regimes (2007). However, the cases that he analyses are all heavily dependent on oil. This book assesses the impact of mineral extraction on both regime stability and social insecurity, thereby addressing another gap in the literature. Specifically, it focuses on the contributions and impacts of both industrial and artisanal mineral extraction.

While oil is particularly prone to enclave economies, mineral resources have a more diverse impact, and bauxite and diamonds offer insights into a greater range of minerals (Diallo, 2017). Hansen (2014) argues that, traditionally,

> extractive foreign direct investment (FDI) in extractives (i.e. mining and oil/gas) in Africa has been seen as the enclave economy par excellence, moving in with fully integrated value chains, extracting resources and exporting them as commodities having virtually no linkages to the local economy.
>
> (5)

Although this remains true with oil extraction, in the case of mineral extraction FDIs also have a direct link with the peripheral local economy as the process uses both foreign and local expertise. In the case of bauxite mining in Guinea, enclave benefits are in some degree extended to the areas where bauxite mining activities take place or to areas that are impacted by mining activities. The presence of bauxite resources have resulted in a situation where different social contracts have emerged. These social contracts, as discussed earlier, have contributed to maintaining the coexistence of regime stability and social insecurity. So far, no work has looked at how bauxite mining influences the concept of the traditional social contract between the state and its citizens.

The prominence of the resource curse arguments has led the international community to suggest that addressing governance issues in natural resource–rich countries in Africa should be a priority for stakeholders and especially for donors.

2.7   Governance in mineral resource–rich countries in Africa

Historically, shortcomings in governance have been identified as the primary impediment to socio-economic development in Africa, to the point that it "has become the common explanation for the failure of African development after 30 years of external assistance" (FCO, 2007, online). There are various definitions of governance, which can be classified in three ways: institutions, networks theory and corporate governance. To different academic authors, institutions such as the World Bank have a definition of governance that is commonly used in the African context. The governance literature is vast, and any exhaustive discourse is beyond the scope of this chapter; however, I do summarise the different schools of thought on governance in the following paragraphs.

The new institutionalism school suggests that institutions construct political and administrative behaviours; hence, they are crucial for political governance and for providing "secure property rights" within societies (Acemoglu and Robinson, 2008; Acemoglu, Johnson, & Robinson, 2001; Ostrom, 1990). The second school of thought focuses on network theories, which argue that governance occurs in networks and consists of cooperation for successful realisation of policies (Kickert, Klijn, & Koppenjan, 1997). Third, the corporate governance school expands accountability and governance to suggest that shareholders are not the sole members of an organisation with interest in the management and output of the company (Kay & Silberston, 1995) – corporations are also made accountable to the stakeholders (Demb & Neubauer, 1992). What this third school suggests is that, apart from shareholders, companies now need to meet expectations from clients and suppliers and ensure that their company has a good image (Kay & Silberston, 1995). Because of these different schools, it is possible to find various definitions of governance. For this book, governance is situated within the institutions school of thought.

Governance can also be defined within two frameworks: the first is concerned with "the rules of conducting public affairs" (Mukamunana, 2008:74), and the second "sees governance as an activity of managing and controlling public affairs" (Hyden & Court, 2002:14). While academics usually use the first definition, international actors advocate the second. For this book, governance is taken to mean "the institutional capability of public organizations to provide the public and other goods demanded by country's citizens or their representatives in an effective, transparent, impartial and accountable manner, subject to resource constraint" (World Bank, 2000:48). The capacity of governance to be effective as defined in this case is affected by different challenges.

Some authors believe that neopatrimonialism limits structural reforms and undermines the principles of good and effective governance (Van de Walle, 2001; Médard, 2006; Bayart, Ellis & Hibou, 1999; Chabal & Daloz, 1999; Bratton & Van de Walle, 1997). Leaders have often driven governance reforms to their advantage, and donors have played a role in sustaining neopatrimonial regimes in their liberal reforms (Bayart, Ellis, & Hibou, 1999; Bratton & Van de Walle, 1997; Chabal & Daloz, 1999; Van de Walle, 2001). The shortcoming in the effective implementation of governance initiatives can be blamed both on the unwillingness of African leaders to implement adequate reforms and on donors' inability to implement stringent measures on governments.

Governance of the natural resource sector, in particular, has become a major challenge for most resource-rich countries in Africa. The poor governance of the extractive sector has been linked to various conflicts in African countries, including the DRC, Sierra Leone, Nigeria (Niger Delta) and Liberia. Natural resource governance is taken to mean the following:

> All internal and external considerations that come to play in the management of natural resources including domestic laws, constitutional provisions, cultural practices, customary laws, neopatrimonial practices, and international practices and obligations that govern issues such as ownership, management, extraction, revenue and grievances over natural resources.
>
> (Alao, 2007:31)

As illustrated by Alao (2007), natural resource governance can cover a range of issues. To understand the impact of natural resource extraction, it is useful to take into consideration most of if not all the elements listed. Most mineral extracting societies in Africa face issues of political, economic, structural and social challenges, including poverty, exclusion and inequality, in the social relationship between citizens and the government.

A common agreement amongst international actors is that good governance should comprise the following key elements: accountability, transparency, participation, fighting corruption and an effective legal and judicial framework (UNDP, 1997; World Bank, 1989). The international community is confident that, if adopted, good governance initiatives would improve the extractive sector in Africa. Poor governance constitutes the opposite of good governance, including non-transparency, unaccountability, lack of participation and corrupt institutions (Moore, 2001). Recently there have also been various debates on how natural resources governance contributes to stability.

According to Alao (2007:276), "people go to war over natural resources when structures to manage these resources and the people to supervise the management of these structures fall below expectation and need". This implies that the governance of the natural resource sector has the potential to impact both stability and instability in resource-rich countries in Africa depending on how resources are managed, how revenues from resources are spent and where they are spent.

In recent years, international institutions have become the guardians of good governance, and many have made it a mission to improve governance in Africa, specifically in resource-rich countries. The governance initiatives are driven by the international community, including the United Nation Agencies, the World Bank, public and private sector agencies, regional and national governments and international NGOs (Alley et al., 2007; Dingwerth & Pattberg, 2006; Weiss, 2000). The involvement of various actors who are working towards improving governance across the world has led to the emergence of the concept of global governance, which, Alao (2007) explains,

> Is a process whereby external institutions attempt to develop broad policies on governance and try to impose these on countries especially developing ones. In a way, the idea of global governance came as a means of instituting some moral codes to the activities of nations without fundamentally altering the structures governing international relations or the sovereign status of states.
>
> (273)

So far, the outcomes of global governance initiatives have not always led to positive outcomes (Norman, 2012). As a result, how global governance initiatives have been implemented has received criticisms. First, although these initiatives are promoted primarily in developing countries, large donors – specifically the World Bank and the IMF – lead policy design and implementation. Second, "decisions are made and policies implemented by leading industrialised countries (e.g. the G7) because they represent the largest donors without much consultation with poor and developing countries" (Bretton Woods Project, 2005, online). According to Ferguson and Lohmann (1994), agencies such as the World Bank, the IMF and national governments are not the type of social actors likely to advance the empowerment of the poor. They suggest that proposed reforms fail because they are prescribed by recognised Western-educated experts who are believed to have solutions to African problems based on their studies of Africa rather than local knowledge or lived experiences. For instance, "toiling miners and abandoned old women know the tactics proper to their situation far better than any expert does" (181). Yet the expert is the one who decides what is best for the miners.

Consequently, the challenge of implementing good governance initiatives in Africa can be linked to two major issues: the first is the ineffectiveness of existing institutions (Alao, 2007; Médard, 2006; Bratton & Van de Walle, 1997); the second is the failure of international organisations such as the UN, the World Bank and transnational bodies to recognise contextual realities of specific countries (Ferguson & Lohmann, 1994). International institutions offer solutions that in practice are not adequate to local realities. The next section will assess the governance initiatives being promoted by transnational organisations in the extractive sector and their impacts on improving accountability, transparency, participation and fighting corruption in the extractive sector.

2.7.1 Role and impact of transnational organisations in Africa

Historically, the African extractive sector has been faced by three major challenges: first, its linkage to conflict; second, the inability of the sector to improve the living conditions of the majority of the citizenry because of poor governance and corruption; and third, the role of the artisanal diamond mining sector as a source of revenue for local communities – and for rebels during conflict, including the wars in Liberia, Angola and Sierra Leone – has been widely publicised. The role that diamonds played in civil wars in Liberia and Sierra Leone was an important trigger for the seriousness of the challenges presented by the extractive sector, presenting immediate threats to people, regimes and businesses. As a result, different initiatives have been constructed to improve the governance of the extractive sector both globally and especially in Africa to prevent it from sustaining civil war and to find ways in which the sector can generate profit that would have a wider positive impact on local communities.

The key initiatives to improve governance in the extractive industry have included the Kimberly Process Certification Scheme (KPCS), the Extractive Industries Transparency Initiative (EITI), Publish What You Pay (PWYP), Voluntary Principles on Security and Human Rights, the United Nations Global Compact (UN Global Compact), the International Council on Mining and Metals (ICMM) and the Extractive Industries Review (EIR). The schemes that have received more attention in mineral resource–rich countries in Africa have been the KPCS and the EITI. This section will focus on the EITI's work because it covers all extractive industries and includes various stakeholders and countries.

2.7.2   *Contextualising the impact of the EITI*

The objective of Transnational organisations is to ensure that mineral resources do not contribute to the outbreak of conflict, that the extractive sector is transparent and free from corruption and that revenues contribute to socio-economic development in resource-rich countries. The EITI was launched to address these challenges. The EITI "is founded upon principles of transparency, accountability and anticorruption". Supporters of the EITI believe that "good governance is necessary to ensure that royalties from oil, gas, and mining projects are used to foster economic growth and poverty reduction" (Hilson & Maconachie, 2009:474). To Aguilar, Caspary and Seiler (2011:15) "at the national level in many of the compliant countries, one of the benefits of implementation has been an increase in engagement among stakeholders—a strengthening of social capital". Those who pro-mote the EITI believe that the resource curse is in fact attributable to "a lack of transparency and accountability around the payments that companies are making to governments, and the revenues that governments are receiv-ing from those companies" (EITI, 2005:2, cited in Hilson & Maconachie, 2009:474). Although many African countries have become signatories of the EITI, they have not seen improvement in accountability and fighting corrup-tion (Corrigan, 2014).

For instance, countries like Chad, Equatorial Guinea, Sierra Leone, Nige-ria, and Guinea are all signatories of the EITI, yet they remain some of the poorest and most corrupt countries in Africa (TI, 2019; UNDP, 2019). This raises questions on the credibility of the EITI's mechanisms and its rigor. As of 2019, the EITI has 52 compliant countries, including Mali, Guinea, Nige-ria, Chad, Sierra Leone, Liberia, Ivory Coast and the DRC (EITI, 2019). Despite criticisms, a paper by Corrigan (2014) suggests that, in regions with abundant natural resources, the EITI has been able to contribute to transpar-ency in the extractive sector but with little impact on democracy, regime stability and corruption.

On the other hand, the argument that transparency and accountability attract businesses seems untrue. Despite the high rates of corruption, insta-bility, and dictatorships, Western businesses have always been present in resource-rich countries like Guinea, Chad, Angola and Equatorial Guinea (Soares de Oliveira, 2007). In addition, as shown with successful failed states, corruption and poor governance do not stop businesses from engaging in a country's extractive sector (Soares de Oliveira, 2007). Although the EITI is a good initiative on paper and is filled with good intentions, the impact of its implementation remains limited in West Africa because of the nature of the extractive resources, the regimes and the voluntary nature of the scheme.

As stated earlier, governments have no direct incentives in becoming accountable or implementing reforms – irrespective of their records, multinationals will still do business with them, and they remain recognisable by the international community (Soares de Oliveira, 2007). Most African governments are mainly interested in collecting revenues from resources and less committed to being accountable to or empowering their citizens (Ferguson, 2006). Mineral resources have facilitated the ability of resource-rich countries to be unaccountable to citizens, and an attempt to reverse this scenario will take more than few voluntary schemes.

In cases where revenue is generated from taxes, governments feel that they are accountable to taxpayers, and taxpayers in return have the power and incentive to ask for accounts of revenue expenditures (Moore, 2004). When government revenues are generated from mining rather than taxes, it is difficult to ensure accountability and enforce anti-corruption initiatives. In these cases, governments feel more accountable to businesses than to citizens. In countries that rely on the extractive industry sector, the "autonomy of the state from citizens" makes it difficult to implement accountability and transparency initiatives (2004). In conclusion, although several African countries have become signatories of initiatives such as the EITI, governance of the extractive sector has not necessarily changed. While donors are kept happy by compliance with the EITI, more efforts are needed to ensure transparency and accountability within the African mining sector.

2.8 Conclusion and contribution to the academic literature

Few attempts have been undertaken to explain how the nature of resources and modes of extraction sustain stability and instability, how mineral rent contributes to regime stability and instability and how states can fail their citizens yet retain international legitimacy. There appear to be no studies explaining how mineral resources contribute to the coexistence of both regime stability and social insecurity, as has been the case in Guinea. Specifically, there are no studies that explain how bauxite mining has contributed to the presence of different forms of social contracts. Guinea is rich in mineral resources, but it is still affected by poverty, dictatorship, neopatrimonialism and poor economic growth. Despite these unique attributes, the country has managed to maintain the coexistence of regime stability and social insecurity and prevent conflict over the years. This book offers further insight on the conditions and underlying contextual realities that enable resource-rich countries (like Guinea) to maintain the coexistence of both regime stability and social insecurity through the presence of different forms of social contracts facilitated by mining.

References

Abdullah, I., 1998. Bush path to Destruction: The Origin and Character of the Revolutionary United Front, Sierra Leone. *Journal of Modern African Studies.* 36(2), pp. 203–235.

ACAPS, 2014. *Sierra Leone: Country Profile.* [online] ACAPS. Available from: www.acaps.org/img/documents/c-acaps-country-profile-sierra-leone.pdf [Accessed on 29 December 2014].

Acemoglu, and Robinson, J., 2008. *The Role of Institutions in Growth and Development. Commission on Growth and Development. Working paper*; no. 10. Washington, DC: World Bank.

Acemoglu, D., Johnson, S.H. and Robinson, A.J., 2001. An African Success Story: Botswana. In: D. Rodrik, ed. *In Search of Prosperity.* Princeton, NJ: Princeton University Press, pp. 80–119.

Adedeji, A., 1993. *Africa Within the World: Beyond Dispossession and Dependence.* London: Zed Books.

Aguilar, J., Caspary, G. and Seiler, 2011. *Implementing EITI at the Sub National Level: Emerging Experience and Operational Framework.* Washington, DC: World Bank.

Ahluwalia, D. Pal S., 2001. *Politics and Post-Colonial Theory: African Inflections.* London: Psychology Press.

Ake, C., 1975. A Definition of Political Stability. *Comparative Politics.* 7(2), pp. 271–283.

Alao, A., 2007. *Natural Resources and Conflict in Africa: The Tragedy of Endowment.* Rochester, NY: University of Rochester Press.

Alley, P., Bermann, C., Danielson, L., Feldt, H., Mahalingam, S., Nadal, A., Nair, C., Nguiffoand, S. and Siakor, S., 2007. *To Have and Have Not: Resource Governance in Africa in the 21st Century.* A memorandum of the Heinrich Böll Foundation.

Andersen, D., Jørgen, M., Lasse, L.R. and Svend-Erik, S., 2014. State Capacity and Political Regime Stability. *Democratization.* 21(7), pp. 1305–1325.

Auty, Richard M., 1993. *Sustaining Development in Mineral Economies: The Resource Curse Thesis.* London: Routledge.

Ballentine, K. and Nitzschke, H., 2005. *Profiting from Peace: Managing the Resource Dimension of Civil War.* Boulder: Lynne Rienner Publishers Inc.

Bangura, Y., 1997. "Understanding the Political and Cultural Dynamics of the Sierra Leone War: A Critique of Paul Richards's Fighting for the Rainforest". *African Development.* 22(3/4), pp. 117–148.

Bannon, I. and Coller, P., 2003. *Natural Resources and Conflict: What We Can Do. In Natural Resources and Violent Conflict: Options and Actions.* Washington, DC: World Bank.

Bar-Tal, D. and Jacobson, D., 1998. Psychological Perspective on Security. *Applied Psychology: An International Review.* 47(1), pp. 59–71.

Basedau, M., 2005. Contéxt Matters: Rethinking the Resource Curse in Sub-Saharan Africa. *Working Paper Series.* 1, GIGA (GIGA German Institute of Global and Area Studies).

Basedau, M. and Lay, J., 2009. Resource Curse or Rentier Peace? The Ambiguous Effects of Oil Wealth and Oil Dependence on Violent Conflict. *Journal of Peace Research*. 46(6), pp. 757–776.

Basedau, M. and Richter, T., 2011. Why Do Some Oil Exporters Experience Civil War But Others Do Not?: A Qualitative Comparative Analysis of Net Oil-Exporting Countries. *Working Paper Series 157*. Hamburg: GIGA (GIGA German Institute of Global and Area Studies).

Bayart, J.F., 1993. *The State in Africa: The Politics of the Belly*. New York: Longman.

Bayart, J.F., Ellis, S. and Hibou, B., 1999. *The Criminalization of the State in Africa*. Oxford: James Currey.

Boone, C., 2003. *Political Topographies of the African State: Territorial Authority and Institutional Choice*. Cambridge: Cambridge University Press.

Bratton, M. and Van de Walle, N., 1997. *Democratic Experiments in Africa: Regime Transitions in Comparative Perspective*. Cambridge: Cambridge University Press.

Bretton Woods Project, 2005. *What Are the Main Concerns and Criticism about the World Bank and the IMF? FAQ*. [online] Bretton Woods Project. Available from: www.brettonwoodsproject.org/2005/08/art-320869/ [Accessed on 20 June 2013].

British Foreign & Commonwealth Office (FCO), 2007. Governance in Africa: What Does It Mean and Why Does it Matter?. London: FCO.

Cameron, W.B. and McCormick, T.C., 1954. Concepts of Security and Insecurity. *American Journal of Sociology*. 59(6), pp. 556–564.

Chabal, P. and Daloz, J.-P., 1999. *Africa Works: Disorder as Political Instrument*. Bloomington: Indiana University Press.

Chazan, N., Mortimer, R., Ravhenhil, J. and Rothchild, D., 1992. *Politics and Society in Contemporary Africa*. Boulder: Lynne Rienner.

Cliffe, L. and Luckham, R., 1999. Complex Political Emergencies and the State. *Third World Quarterly*. 20(1), pp. 27–50.

Collier, P., 2000. Doing Well Out of War: An Economic Perspective. In: Berdal, M. and Malone, D., ed.*Greed and Grievance: Economic Agendas in Civil Wars*. Boulder: Lynne Rienner.

Collier, P. and Hoeffler, A., 1998. On Economic Causes of Civil War. *Oxford Economic Paper*. 50(4), pp. 563–573.

Collier, P. and Hoeffler, A., 2000. Greed and Grievance in Civil War. *Center for the Study of African Economies Working Paper*.

Collier, P. and Hoeffler, A., 2004. Greed and Grievance in Civil War. *Oxford Economic Paper*, 56(4), pp. 563–595.

Corrigan, C., 2014. Breaking the Resource Curse: Transparency in the Natural Resource Sector and the Extractive Industries Transparency Initiative. *Resources Policy* (41), pp. 17–30.

Demb, A. and Neubauer, F., 1992. *The Corporate Board: Confronting the Paradoxes*. New York: Oxford University Press.

Diallo, P., 2017. Social Insecurity, Stability and the Politics in West Africa: A Case Study of Artisanal and Small-Scale Diamond Mining in Guinea, 1958–2008. *The Extractive Industries and Society*, 4(3), pp. 489–496, DOI:10.1016/j.exis.2017.04.003.

Dingwerth, K. and Pattberg, P., 2006. Global Governance as a Perspective on World Politics. *Global Governance*. 12(2), pp. 185–203.

Dunning, T., 2008. *Crude Democracy: Natural Resource Wealth and Political Regimes*. New York: Cambridge University Press.

Englebert, P.K. and Dunn, K.C., 2014. *Inside African Politics*. Lynne Rienner Publishers.

Extractive Industries Transparency Initiative (EITI), 2019. *Factsheet, The Global Standard for the Good Governance of Oil, Gas and Mineral Resources*. March 2019 Available from: https://eiti.org/sites/default/files/documents/eiti_factsheet_en_oct2018_0.pdf [Accessed on 29 May 2019].

Ferguson, J. and Lohmann, L., 1994. The Anti-Politics Machine: Development, Depoliticization and Bureaucratic Power in Lesotho. *The Ecologist*. 24(5), pp. 176–181.

Ferguson, J., 2006. *Global Shadows: Africa in the Neoliberal World Order*. Durham, NC: Duke University Press.

Gberie, L., 2005. *Dirty War in West Africa: The RUF and the Destruction of Sierra Leone*. London: Hurst and Co.

Guberek, T., Guzmán, D., Silva, R., Cibelli, K., Asher, J., Weikart, S., Ball, P. and Grossman, W., 2006. *Truth and Myth in Sierra Leone: An Empirical Analysis of the Conflict, 1991–2000*. Human Rights Data Analysis Group.

Hague, R. and Harrop, M., 2013. *Comparative Government and Politics: An Introduction*. Basingstoke: Palgrave Macmillan.

Hanlon, J., 2005. Is the international community helping to recreate the preconditions for war in Sierra Leone? *The Round Table: The Commonwealth Journal of International Affairs*, 94(381) pp. 459–472.

Hansen, M., 2014. From Enclave to Linkage Economies: A Review of the Literature on Linkages between Extractive MNCs and Local Industry in South Sahel Africa. *DIIS Working Paper 2014*. pp. 1–51.

Herbst, J.I., 2000. *States and Power in Africa: Comparative Lessons in Authority and Control*. Princeton, NJ: Princeton University Press.

Hilson, G. and Maconachie, R., 2009a. The Extractive Industries Transparency Initiative: Panacea or White Elephant for Sub-Saharan Africa? In: Richards, J., ed. *Mining, Society and a Sustainable World*. New York: Springer, pp. 469–491.

Hilson, G. and Maconachie, R., 2009b. Good-Governance and the Extractive Industries in Sub-Saharan Africa. *Mineral Processing and Extractive Metallurgy Review: An International Journal*. 30(1), pp. 52–100.

Hoffman, D., 2006. Disagreement: Dissent Politics and the War in Sierra Leone. *Africa Today*. 52(3), p. 3–22.

Humphreys, M., 2005. Natural Resources, Conflict, and Conflict Resolution Uncovering the Mechanisms. *Journal of Conflict Resolution*. 49(4), pp. 508–537.

Hurwitz, L., 1973. Contemporary Approaches to Political Stability. *Comparative Politics*. 5(3), pp. 449–463.

Hyden, G. and Court, J., 2002. World Governance Survey. *Discussion Paper 1*. Tokyo: United Nations University.

Jensen, N. and Wantchekon, L., 2004. Resource Wealth and Political Regimes in Africa. *Comparative Political Studies*. 37(7), pp. 816–841.

Kandeh, J., 1999. Ransoming the State: Elite Origins of Subaltern Terror in Sierra Leone. *Review of African Political Economy.* 26(81), pp. 349–366.

Kay, J. and Silberston, A., 1995. Corporate Governance. *National Institute Economic Review* (153), pp. 84–97.

Keen, D., 1998. The Economic Functions of Violence in Civil Wars (Special Issue). *The Adelphi Papers.* 38, pp. 1–89.

Keen, D., 2005. *Conflict & Collusion in Sierra Leone.* Oxford: James Currey Publishers.

Kickert, W.J.M., Klijn, E.H. and Koppenjan, J.F.M., 1997. *Managing Complex Networks.* London: Sage.

Kowalke, K.E.C., 2009. *Glare of the Diamond, Botswana, Why AIDS Corporate Social Responsibility Initiatives are not Successful?* Master's Thesis, Lund University.

Le Billon, P. 2008. Diamond Wars? Conflict Diamonds and Geographies of Resource Wars. *Annals of the Association of American Geographers. 98*(2), pp. 345–372. Retrieved from http://www.jstor.org/stable/25515125

Le Vine, V.T., 2004. *Politics in Francophone Africa.* Boulder: Lynne Rienner.

Mähler, A., 2009. Oil in Venezuela: Triggering Violence or Ensuring Stability? A Context-Sensitive Analysis of the Ambivalent Impact of Resource Abundance (11 October, 2009). German Institute of Global and Area Studies (GIGA). *Working Paper No 112.*

Médard, J.-F., 2006. Les Paradoxes de la Corruption Institutionnalisée. *Revue internationale de Politique Comparée.* 13(4), pp. 697–710.

Moore, M., 2001. Political Underdevelopment: What Causes 'Bad Governance'? *Public Management Review.* 3(3), pp. 1–34

Moore, M., 2004. Revenues, State Formation, and the Quality of Governance in Developing Countries. *International Political Science Review.* 25(3), pp. 297–319.

Mukamunana, R., 2008. *Challenges of the New Partnership for Africa's Development (NEDAD): a Case Analysis of the African Peer Review Mechanism (APRM).* PhD Thesis, University of Pretoria.

Norman, M.L., 2012. The Challenges of State Building in Resource Rich Nations. *Journal of International Human Rights.* 10(3), pp. 173–190.

Nugent, P., 2004. *Africa Since Independence.* Basingstoke and New York: Palgrave Macmillan.

Nugent, P., 2010. States and Social Contracts in Africa. *New Left Review.* 63, pp. 35–68.

Nugent, P., 2012. *Politics, States and Social Contracts.* Edinburgh: At The University of Edinburgh.

Oliveira, R.M.S.D., 2007. *Oil and Politics in the Gulf of Guinea.* London: C. Hurst & Co.

Ostrom, E., 1990. *Governing the Commons: The Evolution of Institutions of Collective Action.* Cambridge: Cambridge University Press.

Rashid, I., 1997. Subaltern Reactions: Student Radicals and Lumpen Youth in Sierra Leone, 1977–1992. *African Development.* 22(3/4), pp. 19–43.

Reno, W., 1995. *Corruption and State Politics in Sierra Leone.* Cambridge: Cambridge University Press.

Reno, W., 1998. *Warlord Politics and African States*. Boulder: Lynne Rienner.

Reno, W., 2000. Clandestine Economies, Violence and States in Africa. *Journal of International Affairs*. 53(2), pp. 433–459.

Reno, W., 2003. Book Review: Africa Works: The Political Instrumentalization of Disorder, The Criminalization of the State in Africa. *Journal of Asian and African Studies*. 38(1), pp. 90–95.

Richards, P., 1996. *Fighting for the Rain Forest: War, Youth & Resources in Sierra Leone*. International African Institute. Portsmouth: Heinemann.

Richards, P., 2006. Young Men and Gender in War and Post-War Reconstruction: Some Comparative Findings from Liberia and Sierra Leone. In: I. Bannon and Correia, M. ed. *The Other Half of Gender: Men's Issues in Development*, Washington, DC: World Bank, pp. 195–218.

Rivière, C., 1977. *Guinea: The Mobilization of a People*, trans. Virginia Thompson and Richard Adloff. Ithaca, NY and London: Cornell University Press.

Ross, M.L., 1999. The Political Economy of the Resource Curse. *World Politics*. 51(2), pp. 297–322.

Ross, M.L., 2001. Does Oil Hinder Democracy? *World Politics*. 53(3), pp. 325–361.

Ross, M.L., 2004. What Do We Know about Natural Resources and Civil War? *Journal of Peace Research*. 41(3), pp. 337–356.

Sachs, Jeffrey D. and Warner, Andrew M., 1995. *Natural Resource Abundance and Economic Growth*, NBER Working Papers 5398, National Bureau of Economic Research, Inc.

Sachs, J.D. and Warner, A.M., 1997. Sources of Slow Growth in African Economies. *Journal of African Economies*. 6(3), pp. 335–376.

Sachs, J.D. and Warner, A.M., 2001. Natural Resources and Economic Development: The Curse of Natural Resources. *European Economic Review* (45), pp. 827–838.

Snyder, R., 2006. Does Lootable Wealth Breed Disorder? A Political Economy of Extraction Framework. *Comparative Political Studies*. 39(8), pp. 943–968.

Snyder, R. and Bhavnani, R., 2005. Diamonds, Blood, and Taxes A Revenue-Centered Framework for Explaining Political Order. *Journal of Conflict Resolution*. 49(4), pp. 563–597.

Stewart, F., 2008. *Horizontal Inequalities and Conflict: Understanding Group Violence in Multiethnic Societies*. Basingstoke: Palgrave Macmillan.

Szeftel, M., 1998. Misunderstanding African Politics: Corruption & the Governance Agenda. *Review of African Political Economy*. 25(76), pp. 221–240.

Transparency International (TI), 2019. *Corruption Perception Index 2018*. Available from: www.transparency.org/cpi2018 [Accessed on 29 May 2019].

United Nations Development Programme (UNDP), 1997. *Governance for Sustainable Human Development*. New York: Oxford University Press.

United Nations Development Programme (UNDP), 2019. *Human Development Reports*. Available from: http://hdr.undp.org/en/2018-update [Accessed on 29 May 2019].

Van de Walle, N., 2001. *African Economies and the Politics of Permanent Crisis, 1979–1999*. Cambridge: Cambridge University Press.

Van Wyk, D., 2009. *De Beers, Botswana and the control of a Country-Corporate Social Responsibility in the Diamond Mining Industry in Botswana: Policy Gap 5.* The Bench Marks Foundation.

Vornanen, R., Törrönen, M., Niemelä, J.M. and P., 2012. The Conceptualising of Insecurity from the Perspective of Young People. In: Lopez-Varela, A. ed. *Social Sciences and Cultural Studies: Issues of Language, Public Opinion, Education and Welfare.* Rijeka, Croatia: INTECHopen. pp. 281–297.

Wantchekon, L., 2002. Why Do Resource Dependent Countries Have Authoritarian Governments? *Journal of African Finance and Economic Development.* 5(2), pp. 57–77.

Weiss, T.G., 2000. Governance, Good Governance and Global Governance: Conceptual and Actual Challenges. *Third World Quarterly.* 21(5), pp. 795–814.

World Bank, 1989. *From Crisis to Sustainable Development: Africa's Long-Term Perspective.* Washington, DC: World Bank.

World Bank, 2000. *Can Africa Claim the 21st Century?* Washington, DC: World Bank.

Yates, D., 2009. Enhancing the Governance of Africa's Oil Sector. In: *SAIIA Occasional Paper 51,* Johannesburg: SAIIA, South African Institute of International Affairs.

3 Politics and bauxite mining under the Touré regime (1958–1984)

3.1 Introduction: Guinea faces a choice of integration or independence

In 1958, General de Gaulle, who became president of France in 1959, offered French colonies two choices. The first was to join the 'French Community' (Malinga, 1985; Rivière, 1977). By joining the community, countries would control their internal government and receive aid from France, but France would control their foreign policy, defence and economies. The second was that countries be given full independence with no support and total separation from France (Malinga, 1985; Rivière, 1977). Most colonies voted for the first option. In Guinea, however, Sékou Touré, a former trade unionist, organised and led a strong labour movement working for a vote for independence (Schmidt, 2007). On 28 September 1958, Guineans voted 'No' to joining the French Community, and Guinea was granted its independence on 2 October 1958. Touré became the first Guinean president, leading the Parti Démocratique de Guinée (PDG).

France did not welcome Guinea's choice of independence; indeed, Malinga argues that France was determined to ensure that Guinea's independence failed (Malinga, 1985). France initially ruptured official diplomatic and economic relationships with Guinea; most French nationals were recalled home. Aid from France was cut except for funding to improve the alumina port of Fria (Rivière, 1977). However, a few French nationals with local businesses in Guinea and a few French teachers remained in Guinea (Lewin, 2010). France maintained a secret presence in Guinea through the SDECE (Service de Documentation Extérieur et de Contre Espionage) – the French secret service (Lewin, 2010). This service was used to monitor new political and economic activities in Guinea covertly.

From independence onwards, "France's links with Guinea remained for nearly twenty years on a 'love-hate' basis, very similar to that between a child growing into adolescence and its baffled parent" (Corrie, 1988:49). Of all the French ex-colonies, Guinea was the one with the richest mineral and agricultural resources, being a major producer of bananas, coffee and cacao. The relationship was difficult both because France had not anticipated that Guinea would choose full independence and because Touré had not anticipated France's reaction to Guinea's decision.

The rupture of Guinea's relationship with France at independence created two particular challenges for the country's politics and economy: the first was to maintain regime stability, and the second was to improve economic growth without aid from France. Touré's initial emphasis was bolstering internal security and promoting nationalism, with an emphasis on the unity of the Guinean nation (Corrie, 1988). To meet the first concern, regime stability, Touré created a Guinean army to ensure both internal and external security could be maintained. To address the economic challenge, Touré focused on the mining sector and created a new Guinean currency by leaving the West African CFA franc monetary zone controlled by France.

Touré believed that monetary sovereignty was important for Guinea. The Banque de la République de Guinée (BRG) was created in February 1960, and it immediately launched the new currency – the Guinean franc (GNF) on 1 March 1960; at that point, Guinea officially stopped using the CFA (Cournanel, 2012). The exchange rate was 1 CFA to 1 GNF (Cournanel, 2012). From 1972 to 1986, the syli replaced the GNF. Neither the syli nor the GNF had any value outside Guinea – hence the importance of its mineral resources, whose export contributed to the country's foreign exchange reserves. In an attempt to sabotage Guinea's monetary transition, the French used the SDECE to introduce counterfeit CFA across Guinean markets (Cournanel, 2012), making it difficult for Guinea to buy back the money in circulation. This contributed to delays in the transition to the GNF; yet despite these problems, Touré was able to make a full transition to the GNF and later to the syli.

In 1963, new commercial relationships with France were established. These new relationships were short-lived and broke down in 1965 when Touré accused France of participating in a plot against his regime. Guinea's official diplomatic and commercial relationships with France were re-established in 1975 (CADN, 163.PO/1/40; MAE La Courneuve, 9 May 1977).

This chapter examines the evolution of regime stability under Sékou Touré (1958–1984). The first section discusses how Guinea built its armed forces to maintain regime stability. It explains how the armed forces facilitated the emergence of a strong coercive social contract. The strong coercive social contract highlight state–citizen relationships where regimes have resorted to excessive coercion to govern society. In this framework, the state uses coercive measures to control and discipline citizens. It leaves citizens no choice but to obey the authority of the regime to avoid facing repression from state forces. The second section explores the interaction between mining, the economy and politics under Touré. The main discussion focuses on how economic issues were addressed through new economic alliances and how mining contributed to regime stability while failing to ameliorate the ongoing social insecurity in the country.

The chapter then moves on to discuss the evolution of politics and social insecurity in Guinea under the Touré regime. This section explains how Touré's paranoia contributed to social insecurity and the emergence of a coercive social contract. The consequences of social insecurity and the initiatives undertaken to address some of the problems it created are presented. The chapter ends with a conclusion arguing that the availability of natural resources, specifically bauxite, enabled Touré's coercive regime to maintain regime stability in the face of social insecurity. The chapter addresses two central questions: how did the Guinean regimes between 1958 and 2008 use mineral resources as a tool to maintain regime stability? What has been the nature of governance in Guinea's mining sector between 1958 and 1984, specifically that of the mining of bauxite?

3.2   Building the machinery of power (1958–1984)

When Touré became president, the PDG became the supreme power in Guinea controlling all aspect of public, political and economic life. To Corrie (1988:57), "the ideology of the PDG and Touré was essentially a combination of Marxist-Leninist thinking and African socialism". All major decisions came from the PDG, which banned opposition political parties, making Guinea a totalitarian one-party state: "it encompassed all groups and institutions and restricted individual and corporate liberty" (Kaba, 1977:30). The PDG and Touré are used interchangeably throughout this chapter because "the view of the PDG and the president Touré were virtually the same; all Touré's pronouncements were made in the name of the Party while most national policy guidelines and manifestos were written by the president" (Corrie, 1988:57).

To ensure the survival of his regime, Touré needed to increase the internal and external security of Guinea. At the time of Guinea's independence, the Cold War was escalating between powers of the Western Bloc, led by the United States, promoting capitalism, and the Eastern Bloc countries, led by the Soviet Union, promoting communism. There was ongoing political and military tension between the two blocs, each competing to spread its ideology to other countries. Touré already had some socialist beliefs and was keen to embrace socialist economic principles as a development strategy (Sillah & VanDyck, 2010). As a result, the opportunity to support Guinea was quickly seized by the Soviet Union. The support received from the

Soviet Union and Eastern Bloc countries enabled Guinea to build a strong and well-equipped army in the early days of independence.

The Guinean national army was created in November 1958 with personnel recruited and trained by former members of the French army (Bah, 2009). As soon as Guinea established its national army, the Soviet Union and the Eastern Bloc countries donated military and security equipment to Guinea. Guinea had "2,000 soldiers and received donations of arms from the Soviet Union including 6,000 guns, 600 automatic pistols, 600 cases of grenades, 42 radios, 2,000 (shells/bus) for about 105 guns and 10 cases of Bazookas" (MAE La Courneuve, K3-Afrique-Guinée, vol. 3). From February to August 1959, Guinea received support that included "two shiploads of arms free of charge, 18 advisers from Czechoslovakia, a credit of 140 million roubles repayable in 12 years, at 2.5% interest rate; and infrastructure . . . in return for Guinean goods" (Iandolo, 2012:10). Poland donated items such as clothes, shoes, construction materials, food and arms to Guinea; again in 1959, Guinea received military aid – 300 tons of arms from Czechoslovakia (MAE La Courneuve, K3-Afrique-Guinée, vol. 3). Thus, Guinea soon replaced the economic, training and political support previously offered by France, including technical expertise, with Soviet/ Eastern Bloc aid and expertise.

The development of the Guinean army and its amassing of equipment raised concerns in neighbouring countries, but Guinea reassured them, particularly Liberia and Sierra Leone, that its army would not be a threat to their security (MAE La Courneuve, K3-Afrique-Guinée, 1958–1959). France, too, was concerned about Guinea's expanding military power. It tried to sabotage the military aid being provided to Guinea but not succeed. By 1965, a well-equipped national army was established and, divided into three services: the army, the police and the people's militia (Rubiik, 1987:110). Touré continued to expand the military throughout most of his presidency, such that by 1969

> Guinea had a regular force of 4,800 officers and 30,000 people's militia. By 1979, the regular force had shot up to 8,850 with one armoured battalion, four infantry battalions, one Engineer battalion, 30 T-34/54 medium and 10 pt light tanks. The Navy had 350 men, with six Shanghai Patrol Boats and an Air Force of 500 men, with Mig 14s and a paramilitary force of 8,000 men.
>
> (Rubiik, 1987:110)

Throughout the 26 years of the Touré regime, "the army faithfully served the will and tyrannical love of power of the Leader of the PDG – Touré" (Bah, 2009:153). In particular, the people's militia or 'milice populaire' was used as a political tool by the PDG to control both the armed forces and the population. The PDG introduced its members into the military camps to ensure control of the army (Rubiik, 1987). The people's militia outnumbered the total of the other members of the armed forces. The militia "was largely composed of urban unemployed youths who assumed police functions" (106). These young people were well trained and empowered by the PDG, who relied on them for ongoing surveillance and control of both private and public activities across Guinea.

Under the PDG, any action suspected to be against the PDG's ideology resulted in coercive consequences, which ranged from arrest or imprisonment to public hanging. This built intense fear amongst citizens; consequently, the majority of the population obeyed the PDG. Many of those who were opposed to the PDG's ideology went into exile in neighbouring countries, including Senegal, Ivory Coast, Sierra Leone and Liberia. These actions led to the emergence of a strong coercive social contract between the state – in this case the PDG – and Guinean citizens.

Although Touré claimed to be working for national unity, his regime was coercive, as previously described, and it was not unifying. By the early 1960s Guinea had a well-equipped army, giving the PDG a firm grip on power. However, the state lacked expertise in many development areas. The country also needed to build its economy. The next section will discuss the evolution of the Guinean economy under the Touré regime, with specific focus on the mining sector.

3.3   Mining, the economy and politics in Guinea (1958–1984)

Before its independence, Guinea was already known to have the world's largest reserve of bauxite, as well as other mineral resources like diamonds and iron (MAE La Courneuve, 163.PO/1/40). And France itself had already started investing in the Guinean bauxite mining sector (Steele, 1959, MAE La Courneuve). Touré knew that he could rely on the mining sector to build the economy. During the 1960s, although France was no longer directly involved in Guinea's internal affairs and colonial ties had been cut, France's economic interests in Guinea's mining sector continued through its involvement in companies like CBG and Friguia (MAE La Courneuve, 163. PO/1/40; CADN, KDR/ab no 550/83-RG 03–07). France thus maintained both a political and an economic presence in Guinea primarily through its involvement in the bauxite mining industry.

Despite France's investment in bauxite mining, it initially aimed to sabotage and isolate Guinea to prevent it from gaining Western aid or diplomatic support. In part, it is possible to argue that this was due to France not wanting

Guinea to succeed. If Guinea succeeded, it would prove to France's remaining colonies that they also could succeed without France's help, and the colonies were still important for France's economy and imperial power. As a result, to prevent Guinea receiving aid from Western Bloc countries, France branded Touré a communist leader (Muehlenbeck, 2008). By sabotaging Guinea, De Gaulle wanted to prevent Guinea from receiving the additional support it needed to build the country after the France's departure.

At independence, 90 per cent of the Guinean population was illiterate, and the country had only had three university graduates (Gberie, 2001). Hence, the country did not have the highly skilled professionals and specialists who alone could build the country – specialist support was needed. De Gaulle was sure that Guinea would crumble once French aid had been withdrawn, especially since France had been managing all Guinean affairs with little input from Guineans. Guinea's problems in its early days of independence were exacerbated because, in solidarity with France, no Western country was initially willing to support Guinea.

Touré was determined to ensure Guinea's survival, but, during its first years of independence, he needed urgent support from strong external allies and commercial partners. Again, the Soviet Union was willing to support Guinea by helping to develop its economy (and its military, as already discussed). Nikita Khrushchev (first secretary of the Communist Party of the Soviet Union, 1953–1964) was convinced that "once the developing world saw the full economic potential of socialism, it would turn its back on capitalist and colonialist West, and adopt socialism as a way of life" (Iandolo, 2012:4). The collaboration with Guinea was thus seen by the Soviet Union as a potential win-win situation in which the Soviet Union would support Guinea and then use this support to attract other West African countries to embrace a socialist model of development. In return, Guinea would receive aid from the Soviet Union to build its economy.

The difficult political circumstances during the Cold War Era led Touré both to seek other alliances than the Soviet Union and Eastern Bloc countries and to exploit the potential that mineral resources offered for the stability of Guinea. Touré was a nationalist and against imperial domination, so he chose a policy of 'positive neutrality' during the Cold War and built economic relationships with various countries from both sides (Rivière, 1977; Benot, 1984). Thus, Guinea also became close to and received significant aid from the United States, including a loan of "$400 million to develop its bauxite industry" (Bianco, 1994, online).

As the leader of the first country in West Africa that voted not to join the French Community, Touré was seen as a hero in Africa; his supporters on the continent included Presidents Kwame Nkrumah of Ghana and Modibo Keita of Mali (Sillah & VanDyck, 2010). Nkrumah and Touré shared a common interest in that they both wanted to fight colonialism. Ghana, which had gained independence a year earlier than Guinea, loaned Guinea GHS£10 million (GHS£ = Ghanaian pound) to contribute to its economic development (2010). Apart from this support from Ghana, most of the socio-economic relationships built by Touré were a result of external interest in Guinea's mining sector, which is discussed in the next section.

3.3.1   Early years of the bauxite mining industry (1958–1984)

From 1958 to 1984, the Guinean government chose a policy based on eco-nomic isolation for most sectors – but not for mining (NRED/UNDTC, 1988). Throughout Touré's presidency, although he nationalised most indus-tries, mining was the only industry where companies remained in a joint venture agreement with the state. Nonetheless, all aspects of the mining industry were controlled by the PDG in collaboration with its partners in specific companies. The Code of Economic Activities (MMG, 2009; Camp-bell, 2009), which included guidelines managing the processes of economic activity, governed the mining sector.

There was no stand-alone legislation governing the mining sector, except for shared policy objectives included in agreements between the govern-ment and specific investors. These agreements were based on the following objectives set out by the regime (Camara, 1989:4): control of the industry by the government as a public power; creation of foreign exchange revenue sources; job creation, training and skills improvement of Guinean nation-als; regional development; and acquisition of new technologies. The exploi-tation and management of bauxite mining was controlled directly by the state in partnership with foreign investors (MMG, 2009). In Guinea's post-independence era, three bauxite mining companies became a key source of national revenue and employment, ensuring the economic survival of the Touré regime. In the following paragraphs, we look at these three multina-tionals in more detail.

The first company was ACG (Alumina Company of Guinea). Until 2000, ACG was known as FRIGUIA, which was in turn previously known as FRIA. FRIA, which was established in October 1958, extracted bauxite and refined it into alumina. This was the first alumina refinery in Africa. FRIA was initially a private company. In 1973 a joint venture was established, with the Guinea state owning 49 per cent and FRIALCO, a consortium of private actors, owning 51 per cent (MMG, 2009). The company created in the new joint venture became FRIGUIA. This company contributed greatly to Guinea's economy: in the early 1960s it was contributing three-quarters of total exports by value, as well as providing the majority of foreign exchange earnings (Corrie, 1988:72). It also created a major source of employment for

Guineans as well as a training centre for mining engineers (Soumah, 2008). In the 1960s, it was the only company in Guinea that was both extracting bauxite and processing it into alumina. It constituted the only major industrial activity in Guinea, and the majority of mining revenues came from its activities.

By the 1960s, Touré had already built strong alliances with the Soviet Union; Guinea was receiving aid in the form of gifts and loans and signed major trade agreements with the USSR for the exploitation of Guinea's bauxite. This resulted in the creation of the second company to be discussed here: OBK (Office des Bauxites de Kindia). In 1993, OBK changed its name to SBK (Societé des Bauxites de Kindia). And in 2000, its name changed again to CBK (Compagnie des Bauxites de Kindia).

OBK was created in 1969 and began operating in 1972. OBK was a key asset for the development and support of Touré's ideology. Although OBK was a state-owned mining company, major extraction rights were sold to the Soviet Union in advance. In 1993, SBK became a limited company with public participation and open to private participation (MMG, 2005). In 2000, a convention signed with the Guinean state resulted in CBK, which since 2001 has been managed by RUSAL (2005).

The project for the implementation of SBK was fully financed by the Soviet Union, based on an agreement that 90 per cent of the annual production would be exported to the Soviet Union over a period of 30 years, while 10 per cent could be sold on the spot by the Guinean state; in fact, most of this 10 per cent was also sold to the Soviets and Eastern European countries (NRED/UNDTC, 1988). Under this agreement, of the 90 per cent exported to the Soviet Union, 50 per cent was used to pay Guinea's short- and long-term loans from independence and 40 per cent was used for machinery parts, import fees for products from the Soviet Union and payment of Soviet experts (1988). Thus, the bauxite from OBK was used as a source of payment for a variety of services the Touré regime required.

As already mentioned, at independence, Guinea had limited infrastructure and no qualified human resources to exploit its mineral wealth. The country needed external support for the development of its infrastructure, training and exploitation of its minerals. Through OBK and the various trade agreements, it was possible for Guinea to import equipment from the Soviet Union, including tractors, cars, iron products, geological prospection equipment, construction equipment and agricultural equipment, petroleum, wood, chemical, medicines and food products (MAE La Courneuve, K20-Afrique-Guinée). In return, the Soviet Union received agricultural produce, including bananas, coffee and pineapple, as well as bauxite and diamonds (MAE La Courneuve, K20-Afrique-Guinée).

Over time, minerals gradually replaced agricultural products in Guinea's exports to the Soviet Union. Eventually, most commercial exchanges between the two countries were of Soviet goods, technical support and loans in return for mineral resources (specifically, bauxite) (NRED/UNDTC, 1988). Without support from the Soviet Union and the exchanges facilitated by the creation of OBK, Guinea would not have been able to consolidate political and economic power after independence. Bauxite production replaced money as payment for goods received from the Soviet Union.

The third company was the Compagnie des Bauxites de Guinée (CBG). CBG's concession was signed in 1963 between the Republic of Guinea and Harvey Aluminium Co from Delaware, USA. The Harvey Aluminium Group was formed by a consortium of investors, including Alcoa, Alcan, Pechiney, VAW, Comalco, Uralumina, Billiton and Reynolds (MMG, 2009). This company was established as a joint venture between the state (49 per cent) and private actors (51 per cent). CBG contributed to Guinea's bauxite revenue and established commercial relationships with external actors from different countries that sought Guinea's bauxite resources. The case of CBG is discussed in further detail in Chapter 5 of this book.

By the end of the Touré regime in 1984, the country was heavily dependent on mining. While mining revenues at independence provided 25 per cent of export revenues from Guinea, by the 1970s it was 50 per cent; by the 1980s, 95 per cent (CADN, MRNE/J.R.). Their contributions to GDP increased from 4 per cent in the 1960s to 19 per cent in the 1970s to 22 per cent in the 1980s (CADN, MRNE/J.R.). Furthermore, mining remained a key contributor to the state revenues. The contribution of mining to state revenues increased from 12.5 per cent in 1975 to its highest contribution, 83.6 per cent, in 1986 (BCRG, 2015). Bauxite mining in particular became the key contributor to Guinea's finances because it was the most important industrial mining activity in the country.

It is fair to say that, without bauxite deposits, Guinea would not have survived its first years of independence. OBK was the key to enabling Touré to afford some of the country's much-needed infrastructure, expertise and financial aid. FRIGUIA and CBG contributed to Guinea's export revenue and, crucially, to its international recognition as a legitimate state, which was important for Guinea's survival as an independent state. Guinea's mineral wealth has been crucial for the recognition of its sovereignty at independence as well as its ability to build political and diplomatic relationships with Western countries.

Because of the revenues from bauxite mining, the Touré regime did not need to enforce taxation measures; thus, the regime could afford to be detached from citizens and remain uncountable to them. Regarding the mining companies, the regime received revenues from both its own shares and from taxes. Share ownerships are important sources of revenues for countries with state-owned enterprises. In the case of oil resources, "the majority of oil revenue for government comes not through taxes on foreign companies but rather through state-owned companies. . . . Such non-tax revenues can make up substantial amount of government revenues" (Morrison, 2005:3–5). In the case of Guinea, it is fair to argue that the ability of the regime to generate revenues from both taxes and state shares from the bauxite mining sector contributed to regime stability over time, which allowed the dictatorial nature of the Touré regime to emerge and ultimately go unchallenged. As a result, Touré was able to impose his ideological beliefs on citizens and ignore their demands. This led to the ongoing strong coercive social contract, which was maintained during most of Touré's presidency.

Finally, without bauxite, Guinea would not have been able to maintain either its internal or its external regime legitimacy. Without the strategic importance of Guinea's mineral resources, Touré would not have been able to maintain a position of positive neutrality without conditionality from counterparties. Various nations needed bauxite at the time, and they were prepared to deal with Guinea, despite Touré's refusal to allow any external 'interference' in his internal affairs. As a result, Touré had sufficient external resources, which enabled him to maintain regime stability without external interference. This is an example of Bayart's account of extraversion strategies, in which he illustrates the ability of African leaders to use resources to their advantage (Bayart, 2000). In the case of Touré, mining enabled him to pursue his politics both internally and in his relations with the outside world. The next section focuses on politics under Touré and the context of social insecurity that emerged under his regime.

3.4 From independence to social insecurity in Guinea (1958–1984)

Touré's public rhetoric of unity and nationalism was not matched by his political actions. As illustrated earlier with the people's militia, he created a state of fear and insecurity amongst citizens. In national socialist ideology, the PDG linked all societies, unions and community activities to the party.

In rural areas, the formal private sector was eliminated in favour of state farms and cooperative associations; the PDG abolished private schools, the free press and trade unions (Groelsema, Kante & Reintsma, 1994). Women's and youth groups were all included in the regime's structures, and the PDG had a monopoly over all cultural and sporting activities in Guinea. Although the Touré regime was coercive and divisive, it managed to build a circle of supporters and admirers both internally and externally, as discussed in the next section. However, the majority of Guineans saw Touré as a dictator, despite different strategies used to improve his image in later years.

3.4.1   The Guinean administration (1958–1984): a family affair

During the Touré regime, higher ministerial positions were all occupied by Touré's circle of family and friends. His half-brother Ismaël Touré was the number-two figure in the regime, holding important positions across key ministries (Lewin, 2002), including minister of public works; minister of economic development; minister of agriculture, industry and mines; minister of mines; and, finally, at the time of Touré's death in 1984, minister of economy and finance.

Touré's brother-in-law Moussa Diakité held various ministerial positions for 20 years. Other members of his administration included Mamadi Keita, who was linked to Touré through family ties; his nephew, Commander Siaka Touré; his wife, Andree Touré; Sékou Chérif, another relative by marriage; and, finally, Kesso Bah Touré (Africa Confidential, 1984). Other family members such as brothers-in-law and relatives of his wife were also placed in key ministries. In this way, Touré controlled his entire administration.

This nepotism was also evident in favouritism towards Touré's Malinké ethnic group. Between "1958 and 1966, Malinkés formed 33 percent of the total population and represented 40 percent of the administrative staff. Amongst the governors, regional secretaries, heads of central arrondissement, they accounted for 43 to 51%" (Rivière, 1977:217). The few members of his administration who were not from his Malinké ethnic group were Saifoulaye Diallo, who held key positions, including president of the National Assembly, and Lansana Béavogui, who was Guinea's prime minister for 12 years, from 1972 to 1984. However, towards the end of the Touré regime, Diallo was given less strategic positions.

Despite the measures put in place to control the population, with support from the armed forces and the holding of key administrative positions by his close circle of relatives, Touré remained paranoid that France was trying to destabilise his regime. Any protest by citizens demanding political reforms or improvement to their lives was identified as a potential 'plot' supported by the French, and the responses to these actions were severe. Indeed, anyone caught criticising the regime could be imprisoned or executed (Groelsema, Kante, & Reintsma, 1994). This idea of 'permanent plots' increased social insecurity and reinforced the strong coercive social contract that had been sustained by the regime. These issues, which continued until Touré's death, and their consequences will be discussed in the next section.

3.4.2   Paranoia and plots: reinforcing regime stability and social insecurity

In the early days of independence, it was no secret that France was trying to sabotage Touré and was willing to see him fail. In 1960, an attempt by France to organise a plot against Touré from Senegal was uncovered (Lewin, 2002).

However, even before any real plots started, Touré had already shown that he would use radical measures towards anyone who did not align with the PDG's ideology. As early as April 1959, Touré ordered troops to put down a protest led by a veteran of the French army by force, leading to the deaths of 700 people and injury to thousands more in Guéckédou (Kaba 41, 1998). As Touré's paranoia grew, he adopted more coercive measures to maintain regime stability.

Accounts of 'coups d'état' and plots – some real, some imaginary – became a regular feature of Guinean politics from 1960 to 1977 (Rivière, 1977; Jeanjean, 2005; Lewin, 2010). As a result, Touré took extreme measures that resulted in ongoing social insecurity across the country. The 'Perennial Plot', the idea of serial internal and external plots, was used to justify repression and the imprisonment or execution of many innocent victims (Rivière, 1977; Jeanjean, 2005; Lewin, 2010). The level of fear amongst citizens increased, shows of force were used to manipulate the population and people were obliged to obey the PDG to avoid being associated with potential plots. This use of coercion ensured the survival of the PDG regime (Rivière, 1977). Touré's paranoia about plots did not exclude his close friends from the Soviet Union. In "1961, he expelled the Russian ambassador for interfering in the internal affairs of his country, accusing the Soviets of plotting a Marxist revolution" (Bianco, 1994, online).

In November 1970, there was an attempt to overthrow Touré with the support of Portuguese forces from neighbouring Portuguese Guinea (Groelsema, Kante, & Reintsma, 1994). However, after a few days of fighting, Touré's army defeated the Portuguese, who then withdrew from Guinea. This failed plot increased arrests, assassinations and imprisonment of suspects. Many eminent Guineans were executed and some died in the Camp Boiro prison. Between 1970 and 1971, over 12,000 civil servants were arrested and 80,000 Guineans were recorded missing after the 1970 attack (Kaba 41, 1998). Touré might be considered to have already been paranoid, but the failed plot by the Portuguese in 1970 was rationale used to further reinforce security around him and eliminate more people, thus reinforcing the existence of the coercive social contract.

The other effect of this attack and Touré's reaction was to further distance the regime from the West (Groelsema, Kante, & Reintsma, 1994). In his increasing political isolation after the 1970 attack, Touré further bolstered his defense forces, presumably to ensure the stability of his regime. Between 1970 and 1971, the proportion of defense expenditure was increased by over 10 per cent at the expense of sectors such as public health and agriculture (La Guinée Libre, 1974).

The people's militia was the key source of information for the PDG. Touré relied on the militia, together with the close circle of his ministers and his brother Ismaël, to identify individuals who were seen as a potential threat to his regime. Those identified as potential threats were arrested, imprisoned or killed (Bah, 2009). Some were hanged in public places and then shot dead. In the history of Guinea, Ismaël Touré and Siaka Touré remain infamous for their involvement in the arrest of people sent to Camp Boiro (Bah, 2009), the notorious prison where most of those suspected of disobeying the PDG or identified as a potential threat to the regime were jailed and tortured.

Amongst the many intellectuals killed by the PDG was Diallo Telli, the first secretary general of the Organisation of African Unity (OAU). Telli was amongst the most brilliant African diplomats of the twentieth century. He had returned to Guinea to serve his country; however, Touré disliked the international attention bestowed on Telli and accused him of plotting against the regime. He was arrested and imprisoned in Camp Boiro (Diallo, 1983). Despite the torture to which he was submitted, he maintained his innocence until his death.

Others accused of plotting against Touré were hanged, including some of Touré's former ministers, such as Balde Ousmane, governor of the Guinean Central Bank in 1961; Barry III, secretary of state, who was hanged without trial; Magassouba Moriba, secretary of state; and Commissaire Keita Kara (Kaba 41, 1998). Members of the militia used their position to resolve grudges held against members of the society (1998). As a result, laws were not respected and people lived under the domination of the PDG as a dictatorship. The coercive nature of the Touré regime and the poor socio-economic conditions of life in Guinea drove approximately 2 million Guineans into exile from 1958 to 1984 (Kaké, 1987). As a result, over the years, Touré's popularity decreased across Guinea and he felt the need to improve his popularity internally. The steps taken by Touré to achieve this will be discussed in the next section.

3.4.3   *Decreasing popularity: the PDG leadership softens*

As the number of victims of assassination, arbitrary arrest and execution increased and already poor economic conditions deteriorated further, the Touré regime became increasingly unpopular. Touré was obliged to consider alternative strategies to coercion for maintaining the internal legitimacy of his regime, particularly after both the Portuguese attack in 1970 (see section 3.4.2) and a wave of protests organised by women in 1977.

During the first part of his presidency, Touré showed little interest in Islam, despite Guinea being a Muslim-majority country (Aziz, 1985). After the Portuguese attack on Guinea and the responding increase in oppression by his regime, this changed. Touré realised that his popularity had decreased amongst Guineans, and perhaps surprisingly, he started to show more interest in Islam towards the end of 1970 (1985). It can be argued that this was a strategy used by Touré to improve his image and popularity to gain internal legitimacy by means other than the use of force (Levtzion & Pouwels, 2000). As he turned to Islam, he built relationships with various Islamic organisations, promoted Islamic beliefs in his speeches and undertook the pilgrimage to Mecca (Aziz, 1985).

Developing his new-found Islamic stance, Touré built diplomatic and commercial relationships with several Arab or Islamic countries, including Saudi Arabia, Kuwait, the UAE, Iraq, Iran, Tunisia and Libya (Camara, 2007). In return, Guinea received financial and educational support from these countries. Libya and Algeria offered Guineans university scholarships, Saudi Arabia opened the Dar al-Mal al-Islamiyya Bank in Conakry and countries including Libya, Tunisia, Iraq and Iran established businesses in Conakry when private trade was legalised (Camara, 2007). From 1974 to 1982, Guinea received USD$743 million in aid from Arab development funds (Aziz, 1985). Touré's embracing of Islam offered him additional international legitimacy amongst Arab nations and amongst some members of the Guinean population. Although this brought him some popularity, Touré was unable to convince the entire population of his 'change of heart' – many of those who had suffered at the hands of the regime saw this as an opportunistic way to attract Arab funds to Guinea rather than genuine religious belief (1985). The economic conditions in Guinea remained dire, and by 1977, this was affecting the majority of people, in particular women, who were unable to ensure the livelihood of their families. As a result, the popularity of his regime continued to decline.

The ban on privately owned businesses between 1964 to 1975 led to major economic problems, including shortages of food in markets, imported goods and jobs (Nantes, 163.PO/1/40; Paris, 9 April 1977). In the early years of his presidency, Touré had been able to create a strong following amongst women across Guinea; initially, many key members of the PDG at both the national and the local levels were women. By 1977, the lack of imported goods impacted women market traders whose livelihoods depended on sales of these goods. The frustrations of women sellers because of the poor economic conditions and shortages of merchandise in the markets, including food and clothing, were made worse because the economic police often controlled supplies of goods (Jeanjean, 2005). As a result, a protest was organised by women to ask for an improvement of market conditions. This was the second event that caused Touré to reconsider his policies.

Like all other protests, this was initially seen as a plot by Touré, and it resulted in the arrest and imprisoning of more officials, women and even children as young as ten (Jeanjean, 2005). Nonetheless, the protests organised by the women sent a strong message to Touré. Touré knew that women were a powerful section of society – indeed, one that he had empowered. To Touré, this level of discontent was a major threat to his regime, and he realised that he needed to improve the socio-economic life of citizens to prevent further frustrations and disaffection amongst women.

Ultimately, Touré realised he had pushed the coercive nature of his regime to its limit. The women's protest proved to Touré that when pushed to extremes, they would not fear the repressive measures he was likely to take; furthermore, he needed their support to maintain the propaganda of his regime and maintain the few supporters he had outside his armed forces and family members. As a result, in 1978, Touré started a programme of economic liberalisation in Guinea: he re-established diplomatic and commercial relations with France and with neighbouring countries, and he lifted the ban on private enterprise in 1979. This was the beginning of economic liberalisation in Guinea.

Despite Touré's effort to open the country and liberalise the economy, his popularity continued to decrease. Towards the end of his time in office, he even experienced a physical attack. The attack occurred in May 1982, when a grenade was thrown at him by a young person claiming justice for a former minister, Kabassan Keïta, who had been arrested by Touré (Jeanjean, 2005). This attack signalled that the coerciveness of the Touré regime had already sown deep-seated social insecurity and that the population had become frustrated after years of repressions. Despite Touré's efforts to espouse Islam and the causes of women, the people of Guinea appeared to have largely lost faith with their leader.

3.4.4 A dictator falls

Sékou Touré died on 26 March 1984 during heart surgery in Cleveland, Ohio. Touré's death was sudden, and the PDG was not prepared to put forward a leader to succeed him. Lansana Béavogui was the prime minister (1972–1984) and should have been the person to ensure the continuation of the presidency, but he did not have support from other members of the PDG. Three rival candidates were proposed to take Touré's role (Kaké, 1987).

The first person was his half-brother Ismaël Touré, who, as already described, was the number-two person in the regime. The second potential leader was Moussa Diakité, the minister of habitat and urbanism and brother-in-law of the deceased president. The third candidate was Mamadi Keita, minister of education and another brother-in-law of the deceased president. The military exploited the uncertainty created by three PDG rivals to directly seize power in a coup d'état, which led to Lansana Conté becoming president in April 1984.

Despite Guinea's endowment of natural resources, when Touré died, Guinea was amongst the 20 poorest countries in the world, with an average annual GDP of less than USD$300 per capita (Kaké, 1987). The dictatorial nature of the Touré regime, the strong security services he built with support from the Soviet Union and Eastern Bloc of countries, Guinea's mineral resources and his national socialist agenda resulted in 26 years of Touré's autocratic governance. This led to the exile of many members of Guinean society and political activists. Although he maintained regime stability, he also maintained a strong coercive social contract at the expense of social insecurity – the majority of Guineans lived as if they were in prison. Although Touré defiantly said "we prefer freedom in poverty rather than freedom in slavery", by the time of Touré's death, all Guineans had was a taste of poverty and a lack of freedom. At his death, the majority of Guineans were relieved, and new hopes for freedom animated them.

3.5   Conclusion

The Touré regime was full of contradictions. He led the way to independence to offer Guineans a taste of freedom. In his famous speech against colonisation, he praised freedom. Yet after independence, he imposed an authoritarian and dictatorial regime that banned all opposition to the PDG, creating a single-party state. Thus, the freedom he fought for was never implemented during his rule. Furthermore, since Touré had already established a key source of revenue from bauxite companies and support from the strong alliances he built with the Soviet Union and others, he felt free to pursue his ideology, which sowed the seed for social insecurity in Guinea.

Touré relied on the ruling party militias to intimidate and violently repress any suspected opposition to the ideologies of the PDG. Through this strategy, he was able to maintain regime stability in the face of social insecurity. The PDG became the supreme power to be obeyed, and anyone who did not obey faced repressive actions. Ultimately, a strong coercive social contract was maintained where the population conformed to the will of Touré for fear of their own lives and those of their family members.

Therefore, there is a double irony: Touré fought for freedom, yet sustained a strong coercive social contract. His regime was extremely repressive, and Guineans had to flee in exile to seek freedom in other countries because of social insecurity. By the time of Touré's death, 2 million Guineans, including teachers, civil servants, local leaders and business people, were in exile. He left a legacy of fear that lasted for decades after his death.

The availability of bauxite provided Touré with revenue that enables the exchange of goods and services with the Soviet Union. Mineral resources also enabled Guinea to maintain economic relationships with France and the United States. Without mineral resources, the Touré regime would not have survived its first years of independence, because it would not have had a source of income sufficient enough to sustain the Guinean economy. Bauxite gave Guinea the opportunity to gain international legitimacy at independence. This is due to the strategic importance of bauxite, which is the raw material for aluminium, a metal crucial for many industries, including aerospace, automotive, military, construction and engineering. Because of bauxite, the country's legitimacy was recognised, despite France's attempt to sabotage Guinea.

It is fair to say that under Touré, Guinea developed the attributes of a successful failed state in line with Soares de Oliviera's (2007) argument. Despite the regime's violation of human rights and its dictatorial nature, Guinea's mineral wealth enabled it to gain international legitimacy at independence. The strong army built in the early days of independence in addition to revenue generated from the export of mineral resources enabled Touré to pursue his political ideologies and maintain the coexistence of regime stability and social insecurity. Without mineral wealth, it is unlikely that Guinea would have gained international recognition.

Finally, Touré was able to consolidate power and maintain regime stability in parallel with social insecurity in Guinea for 26 years. The stability of his regime was reinforced by the strong coercive social contract sustained between the PDG and Guinean citizens through the support received from the armed forces, mineral resource assets and powerful alliances with countries interested in the bauxite sector. The scene was set under Touré to use mineral resources and Guinea's armed forces to ensure regime stability in the face of social insecurity. This continued under the regime of Lansana Conté and will be discussed in Chapter 4.

References

Africa Confidential, 1984. *Family Feuds*. Vol. 25 No. 4 Feb 15, 1984. Available from: https://webguinee.net/bibliotheque/sekou_toure/feuds.html

Aziz, P., 1985. Ce qu'il fut. Ce qu'il a fait. Ce qu'il Faut Défaire: La Mecque et les pétrodollars. *Editions Jeune Afrique. Collection Plus*.

Bah, T., 2009. *Trente ans de Violence Politique en Guinée 1954–1984*. Paris: L'harmattan.

Banque Centrale de la République de Guinée (BCRG), 2015. *Statistique, Tableau des Opérations Financières de l'Etat (1974–2008)*. Conakry: BCRG.

Bayart, J.F., 2000. Africa in the World: A History of Extraversion. *African Affairs*. 99(395), pp. 217–267.

Benot, Y., 1984. Sékou Touré: Essayer de Comprendre. *Politique africaine*. 14, pp. 121–124.

Bianco, D., 1994. *Touré, Sekou 1922–1984, Contemporary Black Biography*. [online]. Available from: www.encyclopedia.com/topic/Sekou_Toure.aspx [Accessed on 20 June 2013].

Camara, M.S., 2007. Nation Building and the Politics of Islamic Internationalism in Guinea: Toward an Understanding of Muslims-Experience of Globalization in *Africa. Contemporary Islam.* 1(2).

Campbell, B. (ed.), 2009. *Mining in Africa: Regulation and Development.* Ottawa: International Development Research Council (IDRC).

Corrie, E.M., 1988. *Social Development and Social Policy in Guinea: Health and Education 1958–1984.* PhD Thesis. University of Nottingham.

Cournanel, A., 2012. *L'Economie Politique de la Guinée (1958–2010) – Des dictatures Contre le Développement.* Editions L'Harmattan.

Diallo, A., 1983. *La mort de Teli Diallo: Premier Secrétaire Général de l'OUA.* Paris: Karthala.

Gberie, L., 2001. Destabilising Guinea: Diamond, Charles Taylor and the Potential for Wider Humaniterian Catastrophe. *Partnership Africa Canada.* 1.

Groelsema, B., Kante, M. and Reintsma, M., 1994. *Democratic Governance in Guinea: An Assessment.* Associates in Rural Development.

Iandolo, A., 2012. The Rise and Fall of the 'Soviet Model of Development' in West Africa, 1957–64. *Cold War History.* 12(4), pp. 683–704. ISSN 1468–2745.

Jeanjean, M.H., 2005. *Sékou Touré : Un totalitarisme Africain.* Paris: Editions L'Harmattan.

Kaba 41. Lieutnent-Colonel, C., 1998. *Dans la Guinée de Sékou Touré: Cela a Bien eu Lieu. Mémoires Africaines.* Paris: L'Harmattan.

Kaba, L., 1977. Guinean Politics: A Critical Historical Overview. *The Journal of African Modern Studies.* 15(1), pp. 25–45.

Kaké, I.B., 1987. Sékou Touré:Le Héros et le Tyran. *JA Presses Collection Jeune Afrique Livres.* 3, p. 254.

La Guinée Libre, 1974. *Public Finances of Guinea,* 1964–1973 (in millions of Guinean francs. *La Guinée Libre* (9).

Levtzion, N. and Pouwels, L.R., 2000. *The History of Islam in Africa.* Athens: Ohio University Press and Oxford: James Currey.

Lewin, A., 2002. Jacques Foccart et Ahmed Sékou Touré. *Les Cahiers du Centre de Recherches Historiques.* Revue Electronique du CRH.

Lewin, A., 2010. *Ahmed Sékou Touré (1922–1984). Président de la Guinée de 1958 à 1984.* Paris: L'Harmattan.

Malinga, P., 1985. Ahmed Sékou Touré: An African Tragedy. *African Communist* (100), pp. 56–64.

Ministère des Mines et de la Géologie (MMG), 2009. *Présentation du Secteur Minier et Pétrolier Guinéen.* République de Guinée. Conakry: MMG.

Ministry of Mines and Geology (MMG), 2005. *Guinea, Mineral Resources, Bauxite.* Republic of Guinea. Conakry: MMG.

Morrison, K., 2005. *Oil, Revenue, and Regime Stability: The Political Resource Curse Re-Examined.* Paper presented at the annual meeting of the American Political Science Association, Marriott Wardman Park, Omni Shoreham, Washington Hilton, Washington, DC.

Muehlenbeck, P.E., 2008. Kennedy and Touré: A Success in Personal Diplomacy. *Diplomacy & Statecraft.* 19(1), pp. 69–95.

Oliveira, R.M.S.D., 2007. *Oil and Politics in the Gulf of Guinea.* Columbia/Hurst: Columbia University Press.

Rivière, C., 1977. *Guinea: The Mobilization of a People*, trans. Virginia Thompson and Richard Adloff. Ithaca, NY and London: Cornell University Press.

RubiiK, G., 1987. Social origins of the 1984 coup d'Etat in Guinea. *Utafiti*. (9)1, pp. 93–118.

Schmidt, E., 2007. *Cold War and Decolonization in Guinea, 1946–1958*. Ohio University Press.

Sillah, K. and VanDyck, C.K., 2010. Guinea at a Crossroads: Opportunities for a More Robust Civil Society. *WacSeries*. 1(4), pp. 1–27.

Soumah, I., 2008. *The Future of Mining Industry in Guinea*. Paris: L'Harmattan.

Archives

Centre d'Archives Diplomatiques de Nantes, Archives Diplomatiques, France-Diplomatie, Ministère des Affaires étrangères (CADN), France

Conakry-Ambassade Série: Dossier General: Mines et minerais, questions de ressources naturelles. A/S: Exploitation des mines the diamants. GU 83, Mars 1960-Septembre 1965. GU-8–3; 10/12/1960.

Conakry-Ambassade Série: Dossier Thématique, Le secteur minier en Guinée. 1984–1989. KDR/AB no550/83-RG 03–07.

Conakry-Listes des Sociétés Mixte Opérant Dans le Secteur Minier en Guinée, Octobre 1984. AN/ab no 907/84-RG 05–07.

Conakry, MCAC, Dossier Thématique (1984–1989), Guinée: Nouveau tournant pour la coopération entre l'URSS et la Guinée en matière de bauxite.

Walle, T. *CONAKRY, MCAC, Dossier Thematique, 1984–1989*, 118, Nantes; Department of Technical Co-operation for Development, United Nations, Mineral Investment Policies and Training in Guinea, Interregional Advise on Petroleum and Mineral Legislation, NRED/UNDTCD, July 1988.

Ministère des Affaires Etrangères, La Courneuve, (MAE La Courneuve), Paris, France. Direction des Archives, Archives Diplomatiques, France-Diplomatie

Ambassade De France, NY Herald, 5 April 1959, Service de Presse et D'information, GU-8–3; FRIA, Paris.

Direction Afrique Levant (DAL), A/S Textes officiels, GU-8–3, N0311/AL/10 Mai 1961.

- MAE La Courneuve, 163. PO/1/40; CADN, KDR/ab no 550/83-RG 03–07.
- MAE La Courneuve, K20-Afrique-Guinée.
- MAE La Courneuve, K3-Afrique-Guinée, vol. 3.
- MAE La Courneuve, K3-Afrique-Guinée, 1958–1959.
- MAE La Courneuve, 163. PO/1/40.
- Steele, 1959, MAE La Courneuve.

4 Politics and bauxite mining under the Conté regime (1984–2008)

Following the death of Sékou Touré in March 1984, Colonel Lansana Conté took power through a bloodless coup on 3 April. The coup leaders were Conté and Colonel Diarra Traoré, who became Conté's prime minister. They created the CMRN (Comité Militaire de Redressement National), dissolved the PDG and its militia and freed most of the Camp Boiro prisoners. Some former militia members were recruited into the police and army (ICG, 2010). Most administrative members close to Touré, including his prime minister, Béavogui, were arrested and imprisoned. The coup leaders used claims of abuse, racism and violations of human rights from the PDG as a reason for the coup and promised to improve the socio-economic situation of Guineans.

Despite Guinea's resource wealth, Conté inherited an economy in crisis. Guinea was amongst the 20 poorest countries in the world (Kaké, 1987; Sillah & VanDyck, 2010). By 1985, the Guinean economy was concentrated across three key areas: "the mining sector managed by foreign partners, the public sector which in 1985 employed about 84,000 persons, excluding armed forces, and the private sector, composed mainly of subsistence agricultural producers" (Shapouri, 1988:272–273). Between 1981 and 1985, agriculture employed 80 per cent of the population but contributed less than 1 per cent of export earnings, while mining produced 98 per cent of export earnings (273). In 1984, there were few private businesses, average life expectancy in Guinea was 40 years (Kaké, 1987; Syllah and VanDyck, 2010) and the only aspect of the Guinean economy that was still thriving was the mining sector. From 1984, when Conté came to power, to 1991, mining accounted for over half of state revenue. Although mining contributions to state revenue decreased from 1992 to 2008, mining still constituted about 25 per cent for most of those years, remaining an important contribution, all of which came from a single industry – bauxite mining (BCRG, 2015).

4.1   Bringing in the International Monetary Fund

To improve the economic situation he inherited, Conté solicited the help of the IMF and the World Bank. The IMF (International Monetary Fund) assessed the economic situation and identified three areas for improvement (Gilles, 1989). First, agriculture had to become a key component of economic development and employment. Second, production for export across industries, including mining and agriculture, would only increase in a context of price transparency and a liberalised economy. Third, an immediate currency devaluation would help facilitate business transactions. The IMF also recommended that institutional reforms needed to be undertaken with immediate effect, including a reduction in the size of the public sector (Gilles, 1989). Guinea needed international support and eventually agreed to the IMF's recommendations. This led to a broader liberalisation of the Guinean economy under the IMF and World Bank's Structural Adjustment Programmes (SAP), with a particular focus on the liberalisation of the mining sector.

4.1.1   Structural adjustment programmes and the mining sector in Guinea

The IMF's proposals were not specific to Guinea; they were part of the SAP framework being promoted by the World Bank and IMF across developing nations. These institutions believed that SAPs were the solution to the problems of inflation, poverty, fiscal deficit and GDP growth, which many developing countries faced in the 1970s (FAO, 1999; World Bank [Berg Report], 1981). The SAP reforms included a) currency devaluation; b) changes in fiscal, financial and pricing policy; and c) legal, regulatory and institutional reforms (FAO, 1999:1; Berg, 1981). Changes in fiscal, financial and pricing policy included the elimination of subsidies and the removal of tariffs, while institutional reforms included privatisation of government-owned enterprises and the introduction of cost-recovery programmes.

The IMF and the World Bank reforms in Guinea included exchange rate adjustment, agricultural reforms, economic liberalisation, institutional reforms and reforms within the mining sector (Campbell & Clapp, 1995; World Bank, 1986). These initiatives were set up through collaboration between the IMF and the World Bank: the World Bank focused on mining and agricultural reforms, while the IMF was more focused on fiscal and financial reforms (Campbell & Clapp, 1995; World Bank, 1986). Both, however, worked together on the SAP programmes and promoted private enterprise and increasing economic liberalisation in Guinea.

For Conté, the IMF and the World Bank, these propositions were an attractive solution to improve the economic situation in the short term and reinforce the stability of the regime. As a result, Conté agreed to the recommendations made under the SAP. The financial assistance that the SAP brought was attractive to many regimes at that time, including Conté's, which had limited financial resources. In general, the SAP reforms were used by various African states to their advantage – institutional reforms led to reductions in the numbers of civil servants, but state expenditure increased on the presidency and the defence budget (Van De Walle, 2001). This approach allowed many leaders to maintain their regimes while providing for select elites and their patronage network (2001).

The first SAP reforms in Guinea were implemented in 1986. With these reforms, Conté hoped to increase revenue from the mining sector, promote sector investment and improve Guinea's economic situation. However, the Guinean economy continued to decline. In 1995, almost ten years after embracing trade liberalisation reforms recommended by the IMF and the World Bank, results in the mining sector were disappointing (Campbell, 1995).

One of the factors that contributed to the declining revenue from the bauxite mining industry was declining world aluminium prices (Campbell, 1995). While this was due to external factors beyond the control of the IMF or the World Bank, the SAP reforms did not lead to the anticipated improvements in the civil service, agriculture or socio-economic domains. Indeed, between 1992 and 1994, Guinea ranked last in the UNDP's Human Development Index (UNDP, 1994), and GDP decreased, despite expectations that it would increase (Campbell, 1995). This outcome was not unique to Guinea.

The failures of SAP reforms in most African countries, including Guinea, have been attributed to several factors (Lopes, 1999), including the standardisation of the solutions, which ignored local socio-economic and political realities, and the deliberate reduction of the role of the state (Mkandawire & Soludo, 2003; Lopes, 1999). Each country is different: sustainable change cannot happen by ignoring the social and political context of specific countries, and a single solution cannot work for all developing countries.

Despite the failure of the first round of reforms in Guinea, further policies were put in place to liberalise the mining sector, with support from the IMF and the World Bank, and to attract private investment. To Campbell (2009),

> this strategy rested on the hypothesis that the development of Guinea's mineral resources rested, on the one hand, on opening up the sector to

private operators and market forces, and on the other, the withdrawal of public actors and institutions from the mining sector.

(63)

The involvement of the state in mining activities was reduced to 15 per cent for precious minerals (MMG, 2009), which weakened the institutional capacity and motivation of the state to affect key public policies in the mining sector (Campbell, 2009; Bush, 2010). As liberalisation initiatives were furthered, the revenue from the mining sector decreased. To improve the contribution of the mineral resource sector to state revenues, various mining codes were established in Guinea to serve as the legal framework guiding mining activities.

4.1.2   Economic reforms and the emergence of mining codes in Guinea (1986–2008)

Changes in Guinean mining policies started in 1980 with the adoption of the first code of investment, followed in 1983 by the development of a mineral plan funded by the World Bank (Huijbregts & Palut, 2005:14). When Conté came to power, liberalisation of the mining sector was furthered with financial support from the World Bank and the IMF. The first Guinean mining code was adopted in 1986, and a mining policy was established in 1991 that focused on liberalising the mining sector and attracting investors.

In Guinea, there have been four stages to the evolution of mining legislation. From 1958 to 1986 mining contracts were managed company by company, with no common legislative framework, participation of state capital set at 49–50 per cent and no obligation of any state contribution. State participation in the mining project was, however, compulsory, and with no tax incentive, the sector did not attract significant private investment.

In a second phase, Guinea enacted the 1986 Mining Code, which put in place key provisions for granting exploration activities, mining titles, rights and obligations of holders and the relationship between stakeholders (mainly the mining companies and the state). However, there were shortcomings, including a lack of transparency in state–investor relationships and issues of exploitation rights and the lack of an attractive tax regime.

In the third phase, on 30 June 1995, Guinea passed a law (L/95/036/ CTRN) that created the 1995 Mining Code. This update of the 1986 Mining Code was based on advice from the World Bank and the IMF and focused on increasing the role of the private sector in the mining industry and creating an environment attractive enough to increase investor numbers and decreasing state involvement (Diallo, Tall, & Traoré, 2011). Under Touré, where most mining companies were owned in a public and private joint venture, the state had often owned up to 49 per cent of shares. Under the 1995 code, the percentage of the state was reduced to 15 per cent, reflecting the new liberal approach. The state limited its activities to monitoring rather than direct involvement in the management of mining companies.

On average, in the 1980s and early 1990s, the mining sector contributed about 41 per cent of overall state revenues; however, from 2000 to 2008, mining contributed only 24 (BCRG, 2015). To Diallo, Tall and Traoré (2011), "the decreasing contributions made by the mining sector was due to the reduction in taxes implemented to attract private investors, the adoption of the 1995 mining code, and the increased foreign investment revenue in the sector" (9). It became increasingly clear that the IMF and the World Bank reforms implemented in Guinea had failed to improve the Guinean economy, and by 2008, the larger population felt that mining revenue had not benefited them. As a result, the population sought a revision of the 1995 Mining Code. This process started in 2008 and ultimately led to a new version, the 2011 Mining Code, which was adopted in 2011. Its emphasis is discussed later in this chapter.

4.2 The armed forces and the emergence of a limited coercive social contract in Guinea

As a former soldier, Conté relied heavily on the army to maintain regime stability, which encouraged the emergence of a limited coercive social contract.[1] During the first part of his rule, Conté replaced many people in positions of power in government and national enterprises with members of the armed forces. Under Touré, members of the army held 3 per cent of higher positions in government and 2 per cent in national enterprises; in Conté's early years, these figures were 25 per cent and 27 per cent respectively (Charles, 1989:18). In 1990, however, a transition from military to civilian regime began, and opposition parties were legalised for the first time since Touré had banned them. Multi-party elections were held in 1993, limited press freedom was introduced and the economy was liberalised – all based on prescriptions from the World Bank and the IMF.

Under Conté, the "Guinean military comprised 14,000 men made up of the army, the air force, the navy and the gendarmerie . . . the army alone had 9,700 men" (Bangoura, 2012:100). Conté relied heavily on all branches of the military but used support from the army to maintain regime stability, employing a system of "divide and rule" to maintain control across different ranks (103). Throughout the Conté regime, the army had special privileges, including extra food provisions, housing allowances and higher salaries (Charles, 1989; Bangoura, 2012); these benefits were unavailable to other civil servants. Conté also regularly promoted members of the military to senior government positions; however, as we will see later in this chapter, these promotions benefited only a particular group within the army.

There were three groups within Conté's army (Bangoura, 2012): the first comprised senior officials (including generals, colonels and lieutenant colonels) from diverse ethnic groups who had served since the Touré regime. A second group comprised Conté's Susu ethnic group, which he relied on for full support and whom he gave strategic responsibilities. Because Conté trusted them the most, he promoted them to administrative ranks such as governors, 'prefects' and 'sous-prefects' and granted them more benefits. The objective of these promotions was to ensure that Conté had full control of politics throughout the various regions in Guinea. Guinea has 33 prefectures, and by placing members of the army as prefects, Conté was able to control the regions and ensure internal security across the country.

A third group comprised younger officers who had trained abroad in countries such as France, Morocco and the United States (Bangoura, 2012). This group did not benefit from regular promotions and remained dissatisfied; some of them considered overthrowing the regime. Conté was able to maintain control over the armed forces because of the divisions he created. The first group was unable to organise themselves as a single entity, the second group was loyal to him and it closely monitored the third group of younger officers, who as a result were unable to execute any plans they might have had for regime change (Bangoura, 2012). Although these divisions created discontent, Conté managed to remain in control. By 2008, however, junior officers started to organise mutinies, which threatened the stability of the regime.

Throughout the Conté regime, the military was given strategic positions of authority, to ensure Conté's control over and access to all state resources; the mining sector was not exempt from this. In 1995, a National Security Commission with a coordinating officer for mining companies was created. To support this unit, a security focal point composed of gendarmes or soldiers was dispatched to mining companies and mining sites (Diallo, Tall, & Traoré, 2011). A military attaché was nominated under each National Directorate of Mines; these were called 'Attachés Militaires de Liaison' (2011). A new security unit called Brigade Anti-Fraude des Matieres Precieuses (BAFMP) was also created to monitor and reduce fraud within mining and the processing and exporting of precious commodities (gold, diamonds and precious stones).

Conté bought the army's loyalty by appointing its members to strategic positions in his regime and offering them key benefits unavailable to civil servants or ordinary citizens. Thus, both the political and the administrative apparatuses of the state were controlled by the military. In this way, Conté ensured that his patronage network had access to government resources and maintained their loyalty to him (Melly, 2008). As a result, Conté was able to control all political and administrative decisions to his advantage. For instance, after Conté had legalised multi-party elections, his party, PUP (Parti de l'Unité et du Progrès), won elections in 1993, 1998 and 2003. Conté had rigged the polls, but despite discontent about the fairness of the election amongst opposition members and citizens, they were unable to revoke the outcomes and Conté remained in power. The 2003 elections were meant to mark the end of Conté's final term, but in 2001, he changed the national constitution to increase the presidential time from five to seven years.

Conté's ability to rely on the military, particularly the army, underpinned the stability of his regime. Although his regime was not as oppressive as Touré's, it did not allow the people the freedom to make their own choices; Guinean citizens could exercise their freedom provided it did not threaten regime stability. Also, as seen with election results, Conté paid no attention to the views of Guineans or opposition members.

The military budget remained a state secret in Guinea, so it is difficult to access data on the defence spending of the Guinean administration. Indeed, during interviews in the field, an interviewee suggested that, under Conté, the exact amount allocated to defence was often omitted from the state budget to avoid raising concerns from international donors (interview, Anonymous 1, 10 November 2013). Most interviewees said they did not know the real figure of the defence budget. Any published data on the Guinean defence budget needs to be considered with care, as it is unlikely to be accurate.

I was able to access military expenditure data from SIPRI (see Figure 4.1), since these figures were available from external sources, but they might be missing additional expenses that were not made public or available to foreign bodies. The information from SIPRI is likely to be data relating to the purchase of military equipment outside Guinea, thus excluding expenditure made inside the country, including the cost of exclusive advantages and payments made to members of the army. Nonetheless, this data is useful because it can be taken as an indicator of a lower limit for military expenditure by the Guinean government.

Although the Conté regime went through a political transition from a military to a civilian regime, in practice very little changed either in the power vested in the military or in the way Conté led the country. His leadership style, which was similar to Touré's but less dependent on the use of coercive measures, was that of a dictator. However, despite Conté's control and use of the army as a source of security and a tool for maintaining stability, his regime faced various challenges that threatened national stability. These threats are discussed in the following section.

4.3   Maintaining regime stability in the face of social insecurity

The Conté regime faced three significant threats. The first threat was military mutinies, the second was the war in neighbouring Liberia and Sierra Leone in the 1990s and the third was increasing inflation from 2002 to the end of his regime in 2008, which led to an increase in poverty across the country.

There were three major mutinies during the Conté regime, both over pay raise claims – one in 1996 and two in 2008 (Bangoura, 2012). Although they threatened regime stability, neither led to armed conflict as Conté still had some loyal forces because of the privileges he provided to the army. In response to the 1996 mutiny, Conté promised to increase salaries in the army, directly addressing the main concern that had led to it. However, as described earlier, Conté promoted only a select group to higher administrative ranks with additional benefits. Grievances developed over unequal treatment amongst army members, leading to two mutinies in 2008. The first took place in May 2008; the second, in June 2008. Dissident junior army officers organised the first mutiny with demands that included better salaries, payment of salary arrears, promotions in rank, subsidies for rice and the sacking of the minister of defence (Arieff & Cook, 2009:40; Freedom House, 2009). This group had been frustrated by the lack of promotion of their members over the years and their disadvantageous position compared to that of senior officers.

After a week of clashes between the mutineers and the presidential guard, Conté agreed to end the disputes and pay "salary arrears of $1,100 to each soldier, sack the Defense Minister, and grant promotions to junior officers" (Arieff & Cook, 2009:40). This was sufficient to end the mutiny. In June, police officers that had witnessed the successful riot by junior army officers staged their own mutiny, demanding an improvement in their salaries and in socio-economic conditions (Arieff & Cook, 2009; Freedom House, 2009). In response, the army led a violent crackdown on the police, suppressing the protests. This gave the army additional power over other forces and highlighted the fragility that had existed in Guinea until 2008.

The second major threat to the Conté regime was the war in neighbouring Liberia and Sierra Leone in the 1990s. This war resulted in two key challenges. First, Guinea became home to over 400,000 refugees from Sierra Leone and Liberia (UNHCR, 2000). Amongst these refugees were a number of Guinean returnees who had gone into exile during the Touré regime. This influx strained an already struggling Guinean economy and population. Although some refugees stayed in camps, others lived amongst the wider community and needed shelter, employment and food, which created competition between refugees and Guineans for scarce resources and work opportunities. The war presented a second significant challenge – a series attacks by rebels from Liberia and Sierra Leone between 1997 and 2001.

It is believed that these attacks were led primarily by RUF (Revolutionary United Front) rebels and a small group of dissidents from the RFDG (Rassemblement des Forces Démocratique de Guinée) (Milner, 2005). The RUF was a rebel group based in Sierra Leone that was set up, financed and supported by Charles Taylor, former president of Liberia and former rebel leader of the NPFL (National Patriotic Front of Liberia). The RFDG is believed to be a small group of dissidents consisting of ex-members of the Guinean army who went into exile following the failed mutiny in 1996 (2005). It is believed that Taylor organised the attacks because of Guinea's involvement in the Liberian war – offering assistance to refugees and allegedly supporting the LURD (Liberians United for Reconciliation and Democracy), Taylor's primary opponent during the Liberian war (Gberie, 2001).

By 2001, the rebel attacks on major towns, including Kindia, Macenta, Nzérékoré, Guékédou and Kissidougou, had resulted in about 1,500 Guinean deaths and 100,000 internally displaced people (IDPs), who ran away from the rebels and their destruction of social infrastructures like hospitals, health centres and private houses (Milner, 2005:150–151). The rebel attacks posed a pivotal challenge to stability and security in Guinea. Guinea had received limited military training support from the United States and France in response to the rebel attacks (ICG, 2003). Although the army was able to contain the rebellion, it came at a high cost to the regime, and military expenditure between 2000 and 2002 doubled. This increase in military spending affected the economy as revenue was allocated to defence at the expense of other sectors (Huijbregts & Palut, 2005:16). As Figure 4.1 shows, the year-on-year percentage increase in military expenditure was highest in 2001 (around 90 per cent versus the previous year). After rebels destroyed Guékédou, Conté increased security measures and troop numbers to combat the rebellion.

While Guinea was fighting the rebellion between 1997 and 2001, the mining sector – especially bauxite mining (and in particular CBG) – remained one of the largest contributors to state export revenues (see Figure 4.2). It is possible to argue that the mining industry contributed to regime stability during the 1997–2001 period. Indeed, without a stable source of revenue, it is unlikely that Guinea could have afforded to increase military expenditure or sustain its troops during this time.

The aftermath of the rebellion was further insecurity and significant socioeconomic challenges across the country. According to the ICG (2010), "the 2000–2001 conflict with Taylor's proxies pushed the country near to bankruptcy, as Conté was obliged to purchase tanks, armoured personnel carriers, helicopters and military equipment. He also recruited and armed volunteers to

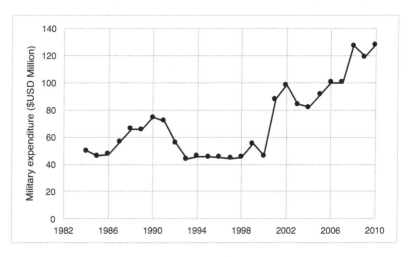

Figure 4.1 Annual military expenditure (1984–2010).

Source: SIPRI, 2015; graph: P. Diallo.

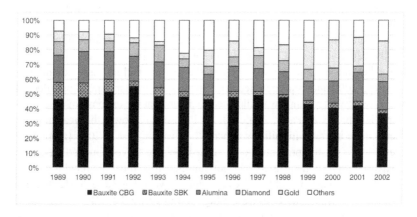

Figure 4.2 Mining contribution to export of goods from 1989 to 2003.

Source: BCRG, 2014; graph: P. Diallo.

fight beside the regular army" (7). Most of the youths recruited to fight the rebel attacks on Guinean soil were "criminally-prone youths and relatives of well-placed senior officers and politicians who had problems in school" (7), and once the rebels were defeated 7,000 youths were left with arms but no employment. The fact that the country was near bankruptcy combined with high numbers of unemployed youths increased the existing social insecurity in Guinea.

Financial constraints resulting from the rebellion prompted the sale of the FRIGUIA (a bauxite and alumina processing plant) to the Russian aluminium company RUSAL at a knockdown price. According to interviews conducted for this study, it was during this post-rebellion period that the government sold FRIGUIA, using funds from this transaction to

> cover the expenditure of the cost of the post-rebellion, the state sold off The FRIGUIA company (Bauxite and Alumina processing plant), which was sold to RUSAL for $22 million. The company's market value was estimated by Ernst and Young in April 2005 at between $22 million to $210 million.
>
> (interview, Guinean mining expert, 15 November 2013)

Prior to the sale of FRIGUIA, Guinea had already signed an agreement that placed CBK (Compagnie des Bauxites de Kindia) under RUSAL's management for 25 years, starting on 21 May 2001. Interviewees from different sectors in Guinea claim that RUSAL gained the right to manage CBK in return for support received from Russia during Guinea's battle against the threats from the rebellion. It is impossible to access data on the Guinean army during the Conté regime because such data was considered a state secret. At the time, the majority of exchanges, including support to the armed forces, were based on bauxite. Therefore, because of the legacy of exchanges and support between Russia and Guinea, we might conclude that Russia was a natural choice for support during Guinea's battle against the rebellion. Although Guinea's bauxite resources enabled Conté to maintain regime stability, this stability did not last long – Conté soon faced a third major threat.

This third major threat to the stability of the Conté regime was increasing inflation and the resulting increase in poverty across the country. As Figure 4.3 shows, inflation increased from 12.9 per cent in 2003 to 31.4 per cent in 2005 to 35 per cent in 2006 (BCRG, 2014).

By 2005, the country was experiencing extreme poverty. In addition to the threats listed in Figure 4.3, Conté's health was deteriorating at this time, and he was losing control of the management of the country, including the armed forces. The president's deteriorating health, the increasing price of goods and services and the rising poverty rate increased discontent amongst citizens who, by 2006, were no longer afraid to stage protests demanding an improvement in their socio-economic situation (ICG, 2003). These protests and the state's response led to increasing social insecurity, which continued until Conté's demise. The protests also led to changes in the Guinean mining codes and to the provision of additional services to communities in bauxite mining areas (cf. Chapter 5).

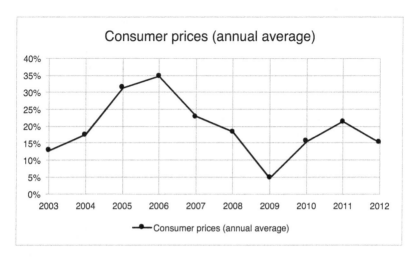

Figure 4.3 Annual inflation indicators (2003–2012).

Source: BCRG, 2014; graph: P. Diallo.

4.4 Social insecurity, popular protest and the limited coercive social contract

Between 2002 and 2006, inflation rose steeply, from 2.3 per cent to 35 per cent (BCRG, 2014). As the inflation rate increased, prices of goods skyrocketed without a corresponding increase in salaries, and the population struggled to afford their usual lifestyles. By 2005, poverty across the country was 53.6 per cent (IMF, 2008); although inflation decreased slightly between 2006 and 2008, it remained high. The economic situation in Guinea did not improve, and the population became frustrated by the state's inability to control the worsening economic crisis.

By 2008, Guinea had been affected by 50 years of dictatorship, extreme poverty, high unemployment and corruption. Inflation coupled with currency depreciation increased social insecurity. As a result, a wave of protests took place from 2004 to 2008 across the whole country, demanding an improvement in the socio-economic situation, which had deteriorated since 2004. The reasons, implications and outcomes of these protests will be discussed in the next section.

4.4.1 Demonstrations at the mines: a protest too far

> The "lack of basic services is primarily the fault of the government, not of mining companies, but the population knows that a way to strike at the government is to halt mining operations".
>
> (civil society member on IRIN, 2008, online)

Although mining communities in Guinea know that it is the responsibility of the state to provide necessary infrastructure, they are also aware that the state will be quicker to acknowledge or respond to their demands if mining companies are jeopardised. Any attacks directed towards mining companies are also an attack on the state revenues. CBG (Compagnie des Bauxites de Guinée), the largest mining company in Guinea, was directly affected by the wave of protests between 2004 and 2008, and this section focuses on the outcome of one of the demonstrations, which took place on the premises of CBG.

The rate of inflation increased and communities across Guinea struggled to survive. "People living near Guinea's mining sites were increasingly taking to the streets to protest the lack of basic services like water and electricity in their communities" (IRIN, 2008, online). These communities were frustrated because, living near mining towns, they knew that it was possible to have access to basic infrastructure as mining companies were already providing this to their employees. As a result, on 31 October 2008, "Boké residents poured into the streets and blocked the railway through which (CBG) transports materials for export" (IRIN, 2008, online). The military was sent from Conakry to Boké to clear protesters from mining sites and re-establish order. The confrontation between the armed forces and the protestors resulted in casualties: "at least two demonstrators were killed when security forces arrived" (IRIN, 2008, online).

Local communities had been determined to continue making demands until they saw an improvement in their lives. Usually, "it is the norm in Guinea – when people stand up for their rights, the reaction is immediate repression. But despite the repression these [mining] communities will continue to claim their rights" (Kabinet Cissé to IRIN, 2008, online). Recognising the strategic importance of mining activities, CBG in 2010 came up with several initiatives to engage local youths and ensure that surrounding communities obtain a more significant benefit from CBG's activities (cf. Chapter 5).

Before these protests, most of the local community support provided by companies active in mining areas such as that run by CBG was geared towards building and improving local infrastructure and funding community development. However, these funds did not seem to have benefited many people, and most local youths lived in poverty despite reforms that had been in place in the mining sector since 1986.

Although the protests took place across the whole country, immediate attention by the state was given to communities in mining areas because mining companies could not afford to stop production. Although the military was used to suppress protests, quick and effective measures were taken by CBG to avoid further disruption. These additional initiatives are discussed in detail in Chapter 5.

4.5   Mining, poverty and protests: a limited coercive social contract

By 2006, the increasing poverty, corruption and unemployment rates furthered social insecurity in Guinea and led to nationwide protests. Despite the existence of a limited coercive social contract, the economic situation in Guinea reached a point where people were not afraid of protesting against the Conté regime. These challenges resulted in various protests across Guinea from 2006 until Conté's death.

Most of the Guinean population is young, with 74 per cent under the age of 35 and 55 per cent below the age of 20 (youthpolicy.org, 2014). In contrast to the situation in neighbouring Sierra Leone and Liberia, where youths played a key role in the civil war, "young people in Guinea for a long time have been key to maintaining social cohesion in a politically volatile and national regional context" (Wybrow & Diallo, 2009:10). However, the ongoing deterioration of life in Guinea pushed youths to contribute to protests related to poverty and dire economic conditions. In "Conakry [in February 2004] rising food prices sent gangs of angry youths into the streets, vandalizing properties, beating shopkeepers, and looting rice trucks" (Jallow, 2005:18).

By 2006, the population was exasperated by poor living conditions. During strikes in February, March and June 2006, Guineans made demands for political change. Guinea's main labour union alliance – the Confédération Nationale des Travailleurs de Guinée (CNTG) and the Union Syndicale des Travailleurs de Guinée (USTG) – launched these historic organised protests, which were proof that the existing limited coercive social contract was threatened.

The CNTG and the USTG demanded increases in wages, union participation in Guinea's socio-economic development and an improvement in the management of the extractive sector. Trade unions had the advantage of being able to mobilise workers from all industries and backgrounds in defence of better living conditions for all; they were also the only groups in Guinea that had a legitimate right to strike without being accused of wanting to destabilise the regime (ICG, 2007). By 2007, union leaders had established themselves as the leading representatives of the Guinean population with support from opposition parties and the National Council of Guinea Civil Society Organisations (CNOSCG). In January 2007, union leaders called a strike to protest against poor governance, corruption, poor socio-economic conditions and high unemployment. There was another strike in February.

The 2007 labour strikes were a consequence of the government's inability to control Guinea's rising inflation and its failure to eliminate poor governance in all sectors, especially mining; concerns over the president's deteriorating health; and a lack of provision for basic social services (IMF, 2008). They were marked by intense and widespread violence across Guinea (ICG, 2007). The strikes paralysed the whole country, including schools, shops and public and private offices. The mining industry, which provided the main revenue for the state, was also affected. The demands made in the 2007 labour strikes included the revision of mining codes to ensure that mining companies' activities benefitted the population as a whole.

A peaceful march on 17 January 2007 in Conakry headed by two key representatives of trade unions, Ibrahima Fofana (USTG) and Rabiatou Serah Diallo (CNTG), was violently repressed by the regime's forces (ICG, 2007). In response, on 18 January 2008, there were uprisings in communities across the country. This time, protestors demanded the departure of Conté. The January uprising was supported nationwide from small towns such as Koundara to neighbourhoods in Conakry such as Kaloum. During these times of crisis, Conté's only support came from the armed forces. This was not enough to curb increasing levels of popular protest, and it became clear to Conté that he faced a significant challenge to his presidency as the whole population was protesting against his regime.

Conté, like Touré, relied on the use of violence and intimidation to suppress protestors and other threats to his regime. Conté responded to the protests in 2007 by ordering security forces to stop the protesters, leading to attacks on civilians by police and armed forces. Clashes between the population and security forces took place 18–23 January 2007 across the country. On 22 January, unarmed protestors in Conakry were brutally attacked by security forces who fired guns on them in an attempt to scatter the crowd.

By February 2007 "the brutal crackdown of the military on civilians resulted in at least 129 dead and over 1,700 wounded, hundreds of them by gunshot" (Human Rights Watch, 2007:6). Civilians were robbed, raped and beaten by the security forces. Children and adults died (Human Rights Watch, 2007). Union offices and private radio stations were attacked. While the regime hoped the people would be cowed, they were ready for change, even at the risk of losing their lives (WRITENET, 2008). Eventually, Conté opened up a space for negotiation between the state and the union leaders representing the population.

The strikes ended in February 2007 in response to a compromise struck between the Conté regime and the broader population represented by the coalition of trade unions and civil society organisations. Conté nominated Lansana Kouyaté as a consensus prime minister, and the government agreed to review mining contracts signed with foreign companies (Campbell, 2009). This led to revisions of the 1995 Mining Code in 2008.

The January 2007 strikes showed that Guineans could affect the governance of the country. The 2007 protests were the first widespread protest in Guinea's history that threatened the stability of its regime and temporarily paralysed the state's activities, including the mining sector. These protests showed that where regime stability is challenged, despite the existence of an ongoing limited coercive social contract, new opportunities for bargaining can be opened. They also proved that increased social insecurity, including extreme poverty and the inability of the state to provide adequate social goods and services to citizens, can lead to the rupture of an existing social contract once people feel they have been left with nothing to protect.

4.5.1   *The promises of a new mining code – a temporary stability*

In the early days of Conte's presidency, the IMF and the government considered the mining sector an essential asset for reducing poverty in Guinea and mitigating socio-economic challenges and unemployment by boosting economic growth, government revenues and basic social services (Campbell, 2009; IMF, 2008). Yet the majority of Guineans experienced no positive impact on their lives, and by the end of 2007, the Guinean population's expectations regarding the mining industry had been dashed.

By 2008, major gaps had been identified in the 1995 Mining Code. Protests and mounting frustration that the mining sector had not yet benefited the majority of the population led to the request from local stakeholders that it be revised. In 2008, negotiations for the revision of the 1995 Mining Code started, involving a group of stakeholders. These stakeholders included officials of the Ministry of Mines and Geology, representatives of civil society organisations, ministerial departments and departments from the chamber of mines, mining companies and donors. This was the first time in Guinea that the full range of stakeholders had been engaged. The significant gaps identified by this group related to local content development, corporate social responsibility, management of mining titles, a lack of focus on measures of good governance, transparency, management of mining revenues and investments and environmental protection in mining areas. Inclusion of these issues in a new mining code formed the basis of the discussions for the revision, which commenced in 2008 and ended in 2011.

Guinea adopted a new mining code in September 2011. It promoted transparency and better socio-economic advantages for Guineans, including a requirement for increased employment of Guineans in the mining sector. There was, however, some disagreement between the mining operators and the government of Guinea regarding the tax regime. To address these issues, on 8 April 2013 an additional amendment to the 2011 Mining Code was adopted by L/2013/053/CNT, a law that added specific provisions to

L/2011/006/CNT. There have been significant efforts to implement the 2011 Mining Code.

While previous mining codes focused on liberalising the mining sector and attracting investors, the 2011 Mining Code emphasised measures to improve transparency and governance in the industry and ensure mining revenues had a positive impact on the socio-economic development of the country and its citizens. While the new concerns taken into account by the latest mining code create space for discussion and opportunities for mining companies to improve their social, economic and environmental impacts on local communities, there remain fundamental challenges regarding its application and implementation.

4.6   Conclusion

Under Conté, the reforms to liberalise the Guinean economic sector and the mining sector in particular did not improve the economy as a whole. By the end of Conté's presidency, poverty was rampant. High inflation and deterioration of purchasing power led to various protests demanding the improvement of the social, economic and political lives of citizens. These protests and demands were suppressed with lethal force, as demonstrators were gunned down and attacked by security forces. By 2007, the number and the intensity of the protests were such that Conté needed to seek a compromise, one acceptable to union leaders and civil society representatives. This marked the beginning of the rupture of the limited coercive social contract that had existed under Conté – protestors proved that they were not afraid of confronting the armed forces to demand improvements in their lives.

Under Touré, trade unions were banned, and there were no independent civil society organisations; hence, there was no space for the organisation of countrywide social protests, particularly given that the militia also monitored the population. Although the liberalisation of the country under Conté did not change the power invested in the military, Conté could not exercise a strong coercive social contract as Touré had. Indeed, he was only able to uphold a limited coercive social contract, and towards the end of his regime, Conté struggled even to sustain that.

Guinea was able to counter rebels' attacks from 1997 to 2001. Apart these, there were no identified external forces trying to sabotage the Conté regime, so there was no basis for imposing the strong restrictions on the population that Touré had. It is possible to argue that revenues from bauxite mining contributed to meeting military expenses and overcoming the economic hardship the country faced in the aftermath of the rebellion, which enabled Conté to maintain political regime stability and protect Guinea's borders and people. Without stable revenue from mining and additional funds from the sale of FRIGUIA, it is unlikely that Guinea would have gathered the financial support needed to fight the rebellion and prevent large-scale conflict within its borders.

Mining resources contributed to regime stability under both the Touré and the Conté regimes. Revenues from bauxite mining provided a stable source of revenue to both regimes (see Chapter 5 for contributions of CBG to state revenues, 1975–2008). Additionally, bauxite mining enabled Touré to maintain regime legitimacy and build political and economic alliances. In the case of Conté, it helped his regime gain additional resources to withstand the rebel attacks between 2000 and 2001 and address some of the economic hardship the regime was facing after the rebellion.

This chapter has explored the interaction between politics, mining, regime stability, the limited coercive social contract and social insecurity under the Conté regime. It has shown that the mining sector can play a key role as a source of revenue and support contributing to regime stability. However, it can also contribute to social insecurity as a result of unmet popular expectations of the sector; this can lead to social protests that can undermine existing social contracts, as was shown in the case of Guinea. Social protests were met with coercive military repression, which only increased social disaffection and led to further clashes between civilians and the armed forces. In the end, however, it opened up a space for bargaining between the state and its citizens that led to stakeholder dialogue and the revision of the mining code. By 2008, however, after 50 years of independence and despite Guinea's mineral wealth, the country remained one of the poorest in the world despite Conté's maintaining regime stability in the face of social insecurity until his death.

Note

1 The limited coercive social contract refers to a situation where citizens have to obey the state but at the same time have some flexibility in their interaction with the it and are free from cruel oppression.

References

Arieff, A. and Cook, N., 2009. *Guinea's 2008 Military Coup and Relations with the United States*, 30 September 2009. CRS Report for Congress. Congressional Research Service.

Bangoura, D., 2012. Gouvernance et réforme du secteur de la sécurité en Guinée. In: Bryden, A. and N'Diaye, B., ed. *Gouvernance du secteur de la Sécurité en Afrique de l'Ouest*. Geneva: Centre pour le Contrôle Démocratique des Forces Armées (DCAF), pp. 99–127.

Banque Centrale de la République de Guinée (BCRG), 2014. *Essentiel Cadrage Macro-économique (Donnée 1989–2008), Contribution du PIB par Secteur (1989-2008)*, Conakry: BCRG.

Banque Centrale de la République de Guinée (BCRG), 2015. *Statistique, Tableau des Opérations Financières de l'Etat (1974–2008)*. Conakry: BCRG.

Bush, R.C., 2010. Mining in Africa: Regulation and Development. *Review of African Political Economy*. 37(126), pp. 547–548.

Campbell, B., 1995. La Banque mondiale et le FMI: Entre la Stabilisation Financière et l'Appui au Développement. *Interventions Economiques* (26), pp. 111–140.

Campbell, B. (ed.), 2009. *Mining in Africa: Regulation and Development*. Ottawa: International Development Research Council (IDRC).

Campbell, B. and Clapp, J., 1995. Guinea's Economic Performance under Structural Adjustment: Importance of Mining and Agriculture. *The Journal of Modern African Studies*. 33(3), pp. 425–449.

Charles, B., 1989. "Quadrillage politique et Administratif des Militaires?". *Politique Africaine* (36), pp. 8–21.

Diallo, C.M., Tall, A. and Traoré, L., 2011. *Les Enjeux de la Gouvernance du Secteur Minier en Guinée*. Conakry: Etude Elaborée et Publiée avec l'Appui de la Coopération Internationale Allemande.

FAO, 1999. *The Effect of Structural Adjustment Programmes on the Delivery of Veterinary Services in Africa*. Animal Production and Health Division. Rome, Italy.

Freedom House, 2009. *Guinea: Freedom in the World*. [online] Freedom House. Available from: https://freedomhouse.org/report/freedom-world/2009/guinea#VbOzUvmCnJc [Accessed on 22 May 2010].

Gberie, L., 2001. Destabilising Guinea: Diamond, Charles Taylor and the Potential for Wider Humaniterian Catastrophe. *Partnership Africa Canada*. 1.

Gilles, L., 1989. Les Institutions de Bretton Woods en République de Guinée. *Politique Africaine* (3), pp. 71–83.

Huijbregts, C. and Palut, J.P., 2005. *Evaluation du Projet FSP*: Decentralisation et Consolidation du Secteur Minier en Guinée. Conakry Service de Coopération D'Action Culturelle à Conakry.

Human Rights Watch (Organization), 2007. *Dying for Change: Brutality and Repression by Guinean Security Forces in Response to a Nationwide Strike*. New York, NY: Human Rights Watch.

Integrated Regional Information Networks (IRIN), 2008. *Guinea: Mining communities will not let up "despite repression"*. [online] IRIN. Available from: http://www.irinnews.org/report/81294/guinea-mining-communities

International Crisis Group (ICG), 2003. *Guinée: Incertitudes autour d'une fin de règn*. [online] ICG. Available from: www.crisisgroup.org/~/media/Files/africa/west-africa/guinea/French%20Translations/Guinea%20Uncertainties%20at%20the%20End%20of%20an%20Era%20French.pdf [Accessed on 20 April 2010].

International Crisis Group (ICG), 2007. *Guinée: Le Changement ou Le Chaos*. [online] ICG. Available from: www.crisisgroup.org/~/media/Files/africa/westafrica/guinea/French%20Translations/Guinea%20Change%20or%20Chaos%20French.pdf [Accessed on 22 November 2014].

International Crisis Group (ICG), 2010. *Guinea: Reforming the Army*. [online] ICG. Available from: www.refworld.org/docid/4c9c68602.html [Accessed on 22 April 2013].

International Monetary Fund (IMF), 2008. *Guinea: Poverty Reduction Strategy Paper, January 11, 2008, Poverty Reduction Strategy Papers (PRSP)*. IMF Country Report No. 08/7. Washington, DC: IMF.

Jallow, B.G., 2005. Guinea: From Democratic Dictatorship to Undemocratic Elections, 1958–2008. In: Saine, A. et al., ed. *Elections and Democratization in West Africa, 1990–2009*. Trenton, NJ: Africa World Press, 2011.

Kaké, I.B., 1987. Sékou Touré:Le Héros et le Tyran. *JA Presses Collection Jeune Afrique Livres*. 3, p. 254.

Lopes, C., 1999. Are Structural Adjustement Programmmes an Adequate Response to Globalization, UNESCO. *International Social Science Journal* (Oxford) (162), pp. 511–519.

Melly, P., 2008. *Guinea: Situation Analysis and Outlook*. [online] WRITENET. Available from: www.ecoi.net/file_upload/1228_1219919131_guinea.pdf [Accessed on 22 November 2014].

Milner, J., 2005. The Militarization and Demilitarization of Refugee Camps in Guinea. In: Florquin, N. and Berman, E.G., ed. *Armed and Aimless: Armed Groups, Guns, and Human Security in the ECOWAS Region*. Geneva: Small Arms Survey, pp. 144–179.

Ministère des Mines et de la Géologie (MMG), 2009. *Presentation du Secteur Minier et Pétrolier Guinéen*. République de Guinée, Conakry: MMG.

Mkandawire, T. and Soludo, C.C., 2003. *African Voices on Structural Adjustment: A Companion to 'Our continent, Our Future*. Trenton, NJ: Africa World Press.

Shapouri, S., 1988. *Global Review of Agricultural Policies*. Economic Research Service, Report No: AGES880304, Washington, DC.

Sillah, K. and VanDyck, C.K., 2010. Guinea at a Crossroads: Opportunities for a More Robust Civil Society. *The West Africa Civil Society Institute (WACSI)*. 4(1).

Sillah, K. and VanDyck, C.K., 2010. Guinea at a Crossroads: Opportunities for a More Robust Civil Society. *WacSeries*. 1(4), pp. 1–27.

United Nations Development Programme (UNDP), 1994. *Human Development Report 1994: New Dimensions of Human Security*. Technical Report. UNDP. New York and Oxford: Oxford University Press.

United Nations High Commissioner for Refugees (UNHCR), 2000. *The State of the World's Refugees 2000*. Geneva, Switzerland: Oxford University Press.

Van de Walle, N., 2001. *African Economies and the Politics of Permanent Crisis, 1979–1999*. Cambridge: Cambridge University Press.

World Bank, 1981 (Berg Report). *Accelerated Development in Sub-Saharan Africa*. Washington, DC: World Bank.

World Bank, 1986. *Guinea: Structural Adjustment Program Project*. Washington, DC: World Bank.

Wybrow, D. and Diallo, P., 2009. *Youth Vulnerability and Exclusion (YOVEX) in Guinea: Key Research Findings*. London: Conflict, Security and Development Group Papers Summary, King's College London.

5 A case study of the Compagnie des Bauxites de Guinée (1958–2008)

5.1 Introduction

The Compagnie des Bauxites de Guinée (CBG) is the largest and amongst the oldest bauxite mining companies in Guinea, a large contributor to state revenue and a provider of basic/social services wherever it operates. CBG was also the largest employer in the industrial mining sector, employing approximately 2,500 national staff per year from 1998 to 2008 (CBG, 2013); CBG's size and scope make conclusions drawn from this case study representative of the industry's contribution to Guinea.

This chapter assesses the impact of industrial bauxite mining on regime stability and social insecurity by examining the socio-economic contribution of CBG in the bauxite mining region of Boké. Additionally, it assesses how an acquiescent permissive social contract allowed CBG to resolve insecurity with limited state intervention in bauxite mining areas. As illustrated in Chapter 2, the acquiescent permissive social contract describes a situation where activities happen as agreed to by the state, under terms and conditions consented to by state and citizens. Sometimes this agreement is informal and is negotiated in return for the stability of mining activities. This explains the relationship between the state, local citizens and industrial bauxite mining companies.

This chapter explains the relationship between the state, local citizens and industrial bauxite mining companies. It begins by introducing CBG. It then moves on to discussing the differences, advantages and disadvantages of life in the bauxite mining enclaves and life outside them. The chapter then discusses the role of CBG as an intermediary between the state and local communities in mining areas, the relationship between the contributions made by CBG in the local communities, community expectations, sources of social insecurity and the emergence of an acquiescent permissive social contract. The chapter then assesses CBG's socio-economic contributions at the local and the national levels, ending with a summary of the main findings.

This chapter answers two questions: one, what has been the impact of industrial mining of bauxite on regime stability and social insecurity in

Guinea?; and two, which type/s of social contract(s) has it contributed to over time?

Social insecurity (see Chapter 2) involves poor socio-economic development including, a) low living standards, b) increasing poverty rates, c) poor development of social infrastructure and d) unemployment that, in extreme cases, leads to social uprising and protests. The foundation of the social contract is based on a promise between the state and citizens in which citizens agree to be governed by the state in return for state-provided development such as welfare and safety. However, in the case of mining areas, mining companies are the ones who provide basic development services to local communities. This has led to the emergence of an acquiescent permissive social contract in bauxite mining areas.

This chapter argues that bauxite mining generates revenues for the state without the state having to deliver services that are ostensibly its responsibility to provide. In effect, bauxite mining companies like CBG become the intermediary between the state and its citizens, as they deliver or expand social goods and services, including health, education, water and electricity, delivering the social contract.

5.2   Compagnie des Bauxites de Guinée (CBG): state-owned enterprise or state proxy?

CBG was created in 1963 as a result of a contract signed between the Guinean government, with a share of 49 per cent, and a consortium of mining companies under Halco Mining Inc., with a share of 51 per cent (that 51 percent was subdivided among three companies – Alcoa had 45 per cent, Rio Tinto Alcan had 45 per cent and Dadco had 10 per cent) (Touré, 2013). While the state is the largest shareowner in CBG, it is a private company that is managed as such. Therefore, CBG and the state are different entities, and this is clear when stakeholders speak about the roles and expectation from CBG and/or the state. The state is an important stakeholder within CBG and, CBG makes important contribution to state revenues. To Mr Touré, former CBG assistant director general, "CBG is a private company whose objective is the exploitation and sale of bauxite, while the Guinean State is the public authority in charge of the management of all the public affairs of the country" (interview, Touré, 16 July 2019). Guinea's government thus has a dual role as a shareholder and as a state with responsibilities to its citizens. But in the bauxite mining region of Boké, the latter role has been largely and increasingly left to CBG over the years. The challenges and consequences of this are discussed throughout this chapter.

Since independence, CBG has produced the highest quantity of bauxite and contributed more to state revenues than any other mining company

(MMG, 2004). This is due to the mining agreement signed with CBG, described as "one of the best agreements in the world as it includes a 65% tax payment on net profit that is given to the government every year" (interview, P. D. Traoré, 7 November 2013; Soumah, 2008). The revenues from CBG have been significant to all governments in Guinea.

CBG's activities are located in the region of Boké, within the district of Sangaredi and Kamsar, in western Guinea. The mine is in Sangaredi, 130 kilometres from Kamsar. Kamsar is home to CBG's bauxite shipping port and processing plant. Once the bauxite is extracted, mining activities are transported by railroad from Sangaredi to Kamsar via several villages, which are thus impacted by CBG's activities. CBG's workers live in Kamsar and Sangaredi, the 'cité' of CBG, where they constitute a majority.

The state, in collaboration with CBG, prioritised infrastructure development to support the new export industry, facilitate the establishment of CBG and provide required services for employees. In 1965, the state established OFAB (Office d'Aménagement de Boké) to lead the planning and building of new local infrastructure, including a port, railroads, a hospital and housing in Kamsar (Soumah, 2008). The state was able to finance these infrastructure projects through loans of USD$74 million (from the World Bank) and USD$25 million (from USAID) as a non-refundable subsidy (Soumah, 2008:114–115). These funds were received between 1965 and 1968.

CBG commissioned additional industrial infrastructure, such as the bauxite mining plant, the pit and equipment for the development of bauxite mining projects. CBG also built housing and social infrastructure in Sangarédi, the second-most important town in the bauxite mining area. In 1995, ANAIM (Agence National d'Aménagement des Infrastructure Miniere) took over the role of OFAB. Under ANAIM, a contract between the state and CBG was signed in which the state leased the infrastructure built by OFAB to CBG at an annual rent of USD$6.5 million, which constitutes an important source of revenue (Soumah, 2008:116). In this way, CBG came to manage the entire relevant infrastructure and could ensure that the buildings, equipment and facilities met its quality standards.

While the infrastructure across CBG's territory met international standards in terms of quality, the majority of infrastructure across Guinea has remained much less developed. The majority of the population had and still has no access to essential services such as water, electricity and adequate health services. The public hospital in Kamsar is known to have the best health care services in all of Guinea. Although the Kamsar hospital was built by the state through OFAB, compared with other state hospitals, CBG provided additional funding to ensure it newer and better equipment and was maintained to a higher standard. CBG also made direct contributions of millions of dollar to the state (see section 5.6, Figure 5.1 and Figure 5.2).

Because of the state's reliance on CBG, it has come to be known as 'the lung of the Guinean economy', implying the economy cannot sustain itself without CBG.

CBG's zones of work have become major destinations for those looking for job opportunities in both the formal and the informal economic sectors. In the early 1960s, there were few people around Boké, Kamsar and Sangarédi; by 1973, "Kamsar had 8,000 people with approximately 3,000 in the town itself and approximately 5,000 in the surrounding area. In 2000, the agglomeration increased to 150,000 inhabitants, almost the 4th or 5th largest city in Guinea" (Soumah, 2008:214). Over the years, the quality of infrastructure and housing built by mining companies such as CBG and the employment opportunities they offer has created internal migration to Kamsar and Sangaredi.

The divide between those who live in CBG housing and those who do not has become a catalyst for high demands by local communities on CBG. In this chapter, 'local communities' are those living in Kamsar, those in proximity to Sangarédi and those directly impacted by CBG's activities along rail transport roads from Sangarédi to Kamsar. Thus, CBG has created mining enclaves in its operational areas and increased internal migration near these. The next section discusses the nature of these enclaves to set the context of the socio-economic impact of CBG and the source of community expectation and social instability in the bauxite mining region of Boké.

5.3 Great expectations: life inside bauxite mining enclaves

Enclaves exist beyond Guinea and are places and economies totally disconnected from the wider socio-economic context of host countries (Ferguson, 2006). Although their nature varies from one country to another in accordance with the nature of the mineral being mined, all share some common characteristics. These similarities are due to support by heavy foreign investment, which is administered by foreign companies in accordance with the norms of their 'home' state – e.g. the United States, France, Australia and Canada; they focus on satisfying an external market and are detached from the wider economy of the country whose minerals they are mining (Olukushi, 2006 in Bond, 2007; Ferguson, 2006; Ferguson, 2005). Kamsar and Sangarédi demonstrate these features, and both have become totally dependent on CBG and completely detached from the rest of the country in terms of infrastructure, standards of living and employment stability.

Outside these types of mining towns, a major informal and formal economy has developed that enables at least some other Guineans to benefit from the mining sector. For instance, outside Kamsar, the largest CBG site, there is a large market with a wide variety of traders. These markets developed to meet the needs of the mineworkers and their families. With their higher salaries, mineworkers have higher purchasing power than that of the majority of Guineans. Hence, trading in the mining areas becomes more profitable to local entrepreneurs than in the rest of Guinea, which contributes to internal migration.

Enclaves represent private gains and public costs. Kamsar and Sangarédi were built to accommodate the production needs of CBG in terms of technology and human resources. Bauxite mining in Guinea is the largest employer after the civil service, and these employees have been able to benefit directly from mining activities through their salaries. The job opportunities offered by the bauxite mining sector are attractive to many Guineans because of the salaries and the lifestyle that it offers to those who live in the mining enclaves.

CBG provides many benefits for its employees, and the standards and the quality of its infrastructure are similar to Western standards. The majority of workers within the cité have access to benefits like free health care, housing and 24/7 water and electricity. Additionally, senior and mid-management employees are provided with additional furniture and household appliances, including refrigerators, stoves and washing machines. However, less than a 15-minute drive from Kamsar and Sangarédi are the realities of life experienced by the majority of Guineans: this division raises frustration amongst local communities who can only dream of the quality of life of the people who live in the enclaves.

5.3.1 A tale of two cities: a widening disparity between those living inside and those living outside bauxite enclaves

The growth of bauxite enclaves has fuelled inequality and growing expectations. The enclaves present major disadvantages that could pose a challenge to the sustainability of mining activities and limit the ability of the mining sector to contribute to socio-economic development within the wider society.

The enclaves often depend completely on mining, and when these activities stop in 30 or 50 years, employees and communities from the host nation are left with no alternative economic opportunities. Perhaps worse, the rest of the population, which generally will not have benefitted much while the mining was ongoing, end up with nothing except the environmental damage left by the mining, such as land degradation and pollution. Such disadvantages can be seen in the case of the town of Fria when mining activities stopped.[1]

As suggested by Ferguson (2005), these economic enclaves offer marginal benefit to the wider society, and Guinea is no exception. Despite recent initiatives by CBG to improve the socio-economic conditions of surrounding communities, life in the enclaves remains a reality that a majority of Guineans want the mining sector to provide to the wider society. The benefits of the enclave are available to those fortunate enough to secure highly skilled mining jobs or managerial roles and to expats working for the mining companies. The division between enclaves and the population outside has increased conflict between local communities and the mining companies.

In the context of CBG, communities living outside enclaves have demanded living standards similar to those provided to CBG workers. When people move to Kamsar and Sangaredi, they expect a higher standard of living and may have unrealistic expectations of job opportunities (e.g. limited employment opportunities and too many people to fill them, especially the roles that have lower education requirements). Although the state is a major CBG shareholder, its involvement in Kamsar and Sangarédi is limited to the infrastructure it rents to CBG. The water and electricity services provided by the state are not reliable, but communities expect to receive reliable services like those in the enclaves. Thus, community demand for improved services is directed to CBG.

Communities are aware of the limited engagement of the state in all rural regions in Guinea. As a result, they have resorted to addressing mining companies directly with specific demands for the socio-economic development of their towns and improved services in return for stability of mining activities. Thus, this has led to the emergence of an Acquiescent Permissive Social Contract.

5.4   Support of local communities and the emergence of an acquiescent permissive social contract

> Before the arrival of CBG and the implementation of infrastructure developed by OFAB-CBG, Kamsar Sub-Prefecture was a district, which had, no sign of development, there was no water, or electricity . . . CBG is the one that first offered water and electricity to Kamsar (at the time).
>
> (interview, Anonymous, 14 July 2019)

CBG's socio-economic engagement has steadily grown from simple training and employment provisions when it started activities in 1963 to a wide range of goods and services in and out of the bauxite enclaves. Initially, Guinea signed mining agreements with CBG that focused on limited socio-economic development at the community level. The agreement was mainly focused on tax payment, training and employment opportunities for Guineans (CTRTCM, 2015). CBG's contributions have created a situation in which the social contract between the state and citizens in its mining areas was effectively fulfilled by CBG. CBG became the main facilitator of socio-economic development in its areas of operation. By becoming an informal intermediary, CBG was expected to deliver socio-economic services to communities, with demands increasing over the years. Consequently, the state's delivery of socio-economic goods and services remained minimal in industrial bauxite regions. This created a space for the emergence of an acquiescent permissive social contract, and CBG has become trapped in its intermediary role.

5.4.1 From training to proxy government: the evolution of CBG's contribution to local development engagement

Once is not custom, but in mining areas of Boké, once becomes an entitlement.[2]
(Interview, Anonymous, 25 June 2019)

CBG's community projects constitute the main mechanism through which the company supports socio-economic development. The support offered to communities around its operating area includes the development of schools, roads, health centres, latrines, markets, cultural centres, administrative infrastructure and access to clean water – these remain limited to mining regions.

In regions where CBG is present, the company has financed most of the development infrastructure, although there have been various difficulties. Between 1987 and 2008, CBG used four successive approaches to fund community infrastructure development. First, from 1987 to 1991, in response to political pressure, CBG paid USD$500,000 in three instalments to the 'prefecture' – the highest government administration office in the locality, which was then supposed to use the funds for the construction of infrastructure like schools, health centres and water facilities (CBG, 2008). Between 1987 and 1991 government officials misused the funds disbursed by CBG, and no infrastructure was completed as planned (2008). Therefore, to ensure better use of CBG community development funds, CBG and the government agreed on a new approach that was implemented in 1993.

Between 1993 and 1998, CBG created a consultation committee that included the assistant director of CBG, the 'prefet', the mayor of Boké, the 'sous-prefets', presidents of CRD (Communautés Rurales de Développement-Rural Development Committees) and union leaders from CBG (CBG, 2008). The committee selected projects presented by the community and assessed their implementation within the available budget. The committee was responsible for overseeing the selected projects. Once completed, the infrastructure was transferred to the local community. During this period, there was clear evidence of new infrastructure and socio-economic development.

Starting in 1999, socio-economic development projects were monitored by a new technical committee that included representatives from CBG, the prefect, the sous-prefects and the presidents of CRD (CBG, 2008). This committee was responsible for putting together tenders, selecting private companies for the execution of projects, evaluating the work and ensuring payment was made upon completion. Although more projects were built, local communities had not participated in the development and infrastructure meant for their use. To address this gap, CBG came up with another new approach.

In 2002, CBG's Advisory Board recommended a new partnership between CBG and the PACV (Programme d'Appui aux Communautés Villageoises) to support development in several communities, including Bintimodia in Boké, Missiara and Daramagnaki in Télémélé (CBG, 2008). The objective was to ensure project beneficiaries were fully involved in the management of development programmes, including their identification, financing and implementation. Under this approach, although CBG provided the majority of funding, communities were required to contribute approximately 20 per cent, including 15 per cent in local materials and 5 per cent in funds to provide long-term maintenance of the project. The implementation of these projects was monitored by representatives from the Ministry of Mines and Geology and the rural communities to ensure development was sustainable, to engage local communities, to build capacity and to reduce poverty (2008). Throughout the different approaches from 1987 to 2010, CBG invested in various infrastructural projects, including 480 classrooms, at least 17 health centres, 15 water boreholes, a 600-kilometre road, 15 administrative housing areas, 40 kilometres of power grid for Kamsar village, 15 covered warehouses and local produce markets, a 4 megawatt power plant for Kamsar village and sports stadiums (CBG, 2013, 2014). CBG extended services, including the provision of electricity and water, to the towns and surrounding villages their rail line passed through.

To improve educational standards in the mining region, CBG made direct contributions to the Institut Supérieur des Mines et Géologie de Boké (ISMGB), which is part of the University of Mines and Geology of Boké. For example, from 1987 to 2008, CBG donated over USD$1.15 million to the Mining Institute in Boké, which contributed to the education of thousands of Guinean students (CBG, 2008). The collaboration between ISMGB led to an increase in the admission of women to the university (in 2008, 15 per cent of students were women). The benefits are not entirely in one direction: the university offers training programmes to CBG staff, and the company often hosts interns and students writing their theses on the mining sector.

CBG's contribution to the development of local socio-economic infrastructure contributed to the development of surrounding towns and led to an increasing number of migrants from elsewhere within Guinea moving closer to mining sites and surrounding villages. Additionally, this led to the growth of trade networks and indirect employment servicing the needs of CBG employees. Small businesses, both formal and informal, have been created around the villages surrounding the mining sites. Daily trade of goods and services has become the source of employment for many, and it constitutes a vital source of revenue for informal traders. The increases in trade peripheral to CBG operations mean many people can have an occupation that provides them with a subsistence income. This is an example of para politics – an informal domain where, as stated in Chapter 2, most political exchanges, including the vast majority of informal business, illicit activities and personal politics, take place, often beyond the control of state structures (Le Vine, 2004:305). In this case, through its activities, CBG has contributed to para-politics, which sustains the livelihoods of many Guineans outside the formal sector of employment.

CBG's support for both the formal and the informal sectors through the procurement of products from the informal sector by its employees and provision of services have contributed to Boké being amongst the most developed regions in Guinea. While these initiatives are not part of the agreement between CBG and the state, it seems safe to assume that CBG is aware of the fact that, without the support of the host communities, it could not pursue its mining activities.

Therefore, CBG initiatives contribute to its ability to maintain an ongoing stable relationship with its host communities. Although the state is the biggest single shareholder of CBG, it is more active in collecting taxes from CBG than in planning or making direct contributions to the socio-economic development of the local community. Additionally, it is most concerned with the infrastructure that it leases to CBG through ANAIM. The lack of active engagement of the state in the larger socio-economic development of the region of Boké was mentioned during interviews with CBG staff. Thus, this leaves CBG with the responsibility to address and sustain ongoing contributions for socio-economic development in the region of Boké.

As discussed in the literature review chapter, African leaders strategically choose to be involved in or leave certain rural areas to fend for themselves within the limited available resources. Boone (2003) argues that states are most likely to engage in rural areas with natural resources and organised rural elites. In such cases, the state contributes to the development of local infrastructure, but while Boone's argument stands true, it is important to note that this is only to maximise revenue from the exploitation of natural resources. What Guinea shows is that, despite the absence of local elites, where there are mineral resources – particularly in the case of large-scale mining – the state invests only in areas that will enable it to generate additional revenues.

For instance, the state invested in the development of infrastructures that it rents to CBG. The rent from the infrastructure enables the state to maximise its revenue. On the other hand, the state does not make significant contributions in the bauxite mining areas, because mining companies are already making a significant investment in socio-economic development. Although the investments are not enough to address the socio-economic shortcomings of local communities, they are often more diverse and of better quality than those provided by the state elsewhere. For instance, while the hospital is owned by ANAIM, CBG provides the functioning budget (USD $3.5 million) for the hospital (interview, Dr F. Doumbouya,[3] 20 December 2019, Kamsar). The pharmacy is also maintained by CBG.

Additionally, the state is aware that large-scale mining companies need a stable environment to work and meet production targets. Therefore, mining companies are ready to maintain stability at any cost to avoid disruptions to their activities. The presence of large-scale bauxite mining operation such as CBG, which have been paying taxes to the state and investing in local communities, forms an essential basis for stability in rural areas. However, the maintenance of stability has come at a cost to CBG as it has created ongoing expectations from local communities.

Bauxite mining activities in Guinea happen in proximity to local communities. In the case of CBG, its mines and factory are in separate towns, and the company has no choice but to commute along national roads and run train lines to pass through local towns. While there is a limited state presence through its administrative institutions, in Kamsar and Boké, these state representatives have neither the capacity nor the power to resolve conflict when they arise. Therefore, when communities complain and request further development projects, to avoid clashes and disruption to operations, CBG finds quick solutions to avoid major losses.

CBG's development projects should have made the local communities more compliant and CBG more powerful. However, dependence has weakened CBG because of a sense of community entitlement and being held accountable for the socio-economic development of the mining region. In turn, CBG is forced to respond to community demands to maintain stability; if it does not, they cannot undertake mining operations. The entire infrastructure used by CBG is in proximity to local communities, and communities are aware that they can easily stop all mining activities. Therefore, CBG's infrastructures and operations are vulnerable.

Additionally, communities know that CBG can provide what they need since it is does so for its employees; thus, they expect the same solutions to be extended to the local community. Therefore, CBG has been held accountable for the delivery of the content of the social contract, which should have been maintained by the state. This taking over of the government's responsibility has made CBG vulnerable and encouraged the emergence of an acquiescent permissive social contract, the implications of which are discussed in the next section.

5.4.2 Hard time outside the bauxite enclaves: the breakdown of Guinea's social contract

Before CBG started its activities in the 1960s, the towns of Kamsar and Sangaredi did not exist; communities moved to these areas once CBG started operating in the region. Therefore, CBG was the first to offer access to water and electricity to local communities in Kamsar village and Sangaredi, starting in 1981 (CBG, 2013). By 2008, state water and electricity, services known as SEG (Guinea Water Company) and EDG (Electricity of Guinea), were not present in either Kamsar or Sangaredi. By providing water and electricity services, CBG started to offer communities services that were usually fully offered and controlled by the state in other regions of Guinea. CBG provided communities with water through public water fountains, which sourced water from CBG's water stations (interview, CBG staff, 2019). These fountains were implemented in public areas and made accessible to local communities based on specific opening and closing times. Thus, by providing communities with water and electricity, CBG created the expectation that these services and support would be sustained, and in the case of water and electricity, the communities came to expect the same level as that offered within the mining enclaves, that is, to have electricity without any blackouts or very limited blackouts. Starting in 2004, CBG failed to maintain regular provision of electricity. This created a source of conflict with local communities, as discussed in the following paragraphs.

CBG's significant support to local socio-economic development has resulted in advantages and disadvantages. The local/host communities came to expect services and goods, including water and electricity delivered by CBG, to continue regularly. Local communities expected CBG to fix any problem they could identify, such as unemployment. Thus, local communities' expectations have grown beyond community development budgets. These expectations are beyond the responsibility of CBG. To maintain stability, CBG has to continue providing socio-economic goods and services such as water, electricity, health service facilities, roads and schools to the local communities in its mining areas.

The main advantage for the communities is that CBG's various contributions enabled some development in their area. This was of benefit to CBG since its operations function more efficiently with good infrastructure, and it also kept the host communities content. However, this created an overreliance on CBG and higher expectations of CBG amongst the host communities. When CBG started to invest in local community development projects in 1987, the state informally abdicated its responsibility. The advantage for the state is that it saved its resources, but more importantly, this prevented communities from putting pressure on the state and therefore contributed to regime stability.

CBG's contribution to communities in the mining areas decreased the pressure on the state and to some extent even rendered the state unaccountable to citizens. Informally, the acquiescent permissive social contract created an expectation of CBG regarding the delivery of specific social goods and services to the local community in return for a stable environment in which CBG could continue its activities. Where community expectations were not met, CBG was forced to negotiate with the local community and find solutions to avoid conflict in the region.

The consequences of the informal role played by CBG in delivering the acquiescent permissive social contract are discussed further, with examples, in the following sections. Interestingly, it can be argued that CBG also chose an acquiescent permissive type of relationship with the government, providing both communities and the state with long-term security. On the one hand, CBG provided social goods to local communities regularly; on the other hand, it paid taxes to the state regularly. As a result, CBG catered to both parties (the community and the state) in return for a smooth running of activities.

Despite the contribution of CBG to local development, local communities from mining areas felt that they did not benefit from direct employment opportunities with CBG or from direct revenues that would improve their livelihoods. These communities saw bauxite extracted from their lands daily yet felt they had not benefited from advantages attached to such employment (interview, Sanoh and Touré, 8 November 2013). Most employment within the company chiefly benefitted the highly skilled professionals with backgrounds like engineering, human resource, finance, health, safety and environment, thus excluding the majority of the local community. While some local community members benefited from employment, they did not perceive the same benefits in terms of housing and salaries compared to the skilled professionals. This did little to reduce social insecurity in the mining regions. Furthermore, this created frustration within the communities that were outside the mining enclaves but within the mining regions (interview, Sanoh and Touré, 8 November 2013). The combination of these issues and frustrations resulted in community protests, which started in 2004, in which communities demanded more resources and support from CBG. The following sections discuss the impact of these challenges and their consequences in further detail.

5.5 CBG as an intermediary – negotiating community expectations, social insecurity and social contracts

The rising community expectations, combined with high internal migration, unemployment, increasing poverty and unmet expectations, let to three major community protests, which took place between 2004 and 2008 across CBG's project locations. These protests started at the same time as other protests in different regions of Guinea, at a time when the country faced an economic crisis because of high inflation and the resultant diminishing purchasing power of Guinean currency (cf. Chapter 4). The difference was that the demonstrations in CBG's areas of operations were targeted at CBG directly, not at the state, and on each of these occasions, CBG's activities were destabilised. This was a consequence of CBG becoming the intermediary between the state and citizens, an overreliance on CBG and high expectations fuelled by a poor understanding of the responsibility of mining companies towards local communities and the responsibility of the state. From 2004,[4] the protests became the medium used by communities in the mining areas when they needed to adjust the existing informal acquiescent permissive social contract.

In 2004, the population of Kamsar protested because of the lack of reliable electricity in the district. The population was frustrated because, within CBG's enclave, electricity was accessible 24/7. The protests lasted four days and resulted in the destruction of administrative infrastructure in CBG's compound, destruction of the power plant and threats towards CBG employees. The renovations of administrative housing and infrastructures cost the company USD$50,963, and the protest put mining operations on standby for four days (CBG, 2008).

A second major protest was organised by women – 'la Marée Rouge'. This took place in Kamsar from November to December 2006. It was led by women from local communities outside the mining enclave who came out in mass protest to demand further improvement of their communities' socio-economic conditions. This time, the result was that the mining operations were on standby for two weeks (CBG, 2008). The women specifically wanted access to better health care, clean drinking water, electricity and employment opportunities for their children. They were frustrated that their children had no access to work within CBG.

The third protest happened in January and February 2007, in solidarity with demonstrations that took place across the country (cf. Chapter 4). This last protest took place to demand further improvement of the socio-economic conditions in Guinea and an improvement in the governance of revenues from the Guinean mining sector to create benefits for the populace. All three protests demanded an improvement in the socio-economic conditions of communities outside the mining enclave. Communities living outside the enclaves but affected by CBG's activities expected further investments, including access to clean water, access to work with CBG, access to food markets, support for agricultural development, further training, capacity building and micro-credit initiatives (interview, Sanoh and Touré, 8 November 2013). CBG presented limited opportunities for unskilled youths from host communities with limited academic backgrounds or formal professional education. Although there are roles that require only basic literacy and numeracy skills, such as drivers, many local youths cannot afford a driver's license, which is a prerequisite.

During 2004–2008, the protests led to CBG stopping activities for a total of 15 days. This inactivity led to a loss in revenue for investors, customers and the government. To prevent further protests and attacks on CBG's assets, the provision of electricity to communities outside the mining enclave became a priority. After the protests of 2004 and 2006, CBG committed an annual investment of USD$1.6 million for community development in Kamsar, Boké and Sangarédi, an increase of USD$1.1 million from the previous annual investment of USD$500,000 (CBG, 2008). The USD$1.6 million was invested in the improvement of water and electricity infrastructures. The investment for the power plant was USD$4 million (CBG, 2008). All these initiatives illustrate how CBG has been under pressure to invest in socio-economic development to maintain stability.

Youths living in the areas near the mines felt they did not benefit from mining revenues. Because of their frustrations, between 2007 and 2008, regular protests occurred (interview, Sanoh and Touré, 8 November 2013). In response, Tres Petites Enterprises (TPE) was launched by CBG to promote enterprise amongst young people native to the mining region through access to small loans and mentoring (interview, Sanoh and Touré, 8 November 2013). TPE encourages youths to create small enterprises rather than wait for aid or permanent external support from CBG. It was an attempt to improve community relations and prevent further disruption.

In 2010, CBG supported the creation of nine TPEs, which offer various services, including renovation and construction of houses, sanitation and cleaning, industrial drain cleaning, jam making, sanitation, reforestation quarries and railway works. Eight TPEs were still functioning in 2017, excluding the only female-managed enterprise (CBG documents, 2019). Although TPEs are open to men and women, men lead the majority of enterprises. There is need to encourage and offer women additional support to successfully develop enterprises. In recent years, there have been claims that the TPEs have contributed to a divide amongst local youths – between those who get support from CBG and youths still frustrated by limited access to opportunities offered. While the TPEs provide opportunities, it is not the responsibility of CBG to address youth unemployment, which is an issue across Guinea. However, any solution offered by CBG creates further expectation. Although the TPE was created after 2008, which is beyond the timeframe of the focus of this chapter, I mention it briefly because the different events from 2004 to 2008, including protests, led CBG to think of new initiatives and means of addressing the growing social insecurity in the mining region.

CBG also engaged in a communication campaign to show the direct contribution that it makes to surrounding communities, again, after the timeline of the focus of this chapter. To protect its investment given the state's limited capacities, CBG has had to find solutions to all the issues that have come up in its operating region. CBG pays taxes to the state. Additionally, it is filling the role of the state institutions within its operating areas, investing in socio-economic development in return for stability for its mining operations.

In bauxite mining areas, when communities protest or make demands, in effect this means they want to renegotiate the terms of the existing acquiescent permissive social contract. The population outside the mining enclaves that lives in poverty has developed increasing expectations for socio-economic improvement, which are not met by the state. A shutdown of CBG because of protests affects the plant and the Guinean economy. CBG becomes the natural target for frustration, resulting in protests in the hope of a quick response. Communities understood that protests on mining sites put pressure on the government as it affects the tax and rent revenues to be received from mining companies. In CBG's case, the state receives revenues both as shareholders and as collectors of rent and taxes revenues.

Attacking CBG's mining operations puts pressure on both CBG and the government, encouraging the company to maintain the existing acquiescent permissive social contract. Consequently, this supports regime stability because CBG quickly resolves any demands addressed to the regime from communities across CBG mining areas.

Conflicts affecting CBG's operations present an important risk to the state. Despite the effort made by CBG to address the needs of local communities, the enclave nature of bauxite mining needs further attention to ensure that the revenues from bauxite mining companies contribute to socio-economic development in the broader region, that is, the whole of Guinea. As discussed in Chapter 6, there have been a growing number of bauxite mining companies and increasing bauxite mining activities since 2010. The presence of new bauxite mining companies with limited positive impacts for local communities has led to further protests in the region of Boké, which require more attention from stakeholders to prevent large-scale mineral related conflict.

5.6   CBG and its socio-economic contributions to the state and local communities

Over the years, CBG has made a regular contribution of millions of US dollars to state revenues from its activities (see Figure 5.1 and Figure 5.2). Although the tax contributions from CBG have decreased at times, its contribution has remained significant for Guinea. Consequently, "because of the tax contributions to be paid by CBG, all the regimes have been very supportive of the company while at the same time trying to extract as much as they can from the company" (interview, Théa, 7 November 2013). The decreasing amount of revenues generated from bauxite production in some years is also linked to the world market price for bauxite, which has not been stable over the years.

To Snyder & Bhavnani (2005), revenue is the source of stability of the state, and the chances of state collapse are higher when a state has no revenue – or as the authors put it, "no revenue, no regime" (10). They believe that mineral resources are likely to contribute to stability if they are non-lootable. They contribute to state revenues through taxes, and in return, the state uses these revenues to strengthen security forces and contribute to the state's expenditure on social welfare (Snyder & Bhavnani, 2005:4–5). In the case of Guinea (see Chapters 3 and 4), revenue from mining played a crucial role in the stability of both the coercive Conté and Touré regimes.

The regular income from the bauxite mining industry going directly to the government finances has been a significant asset to regimes in Guinea,

Figure 5.1 CBG's bauxite production and tax contribution to state revenues (1975–1983).

Source: Figure by author; data from CBG (2013).

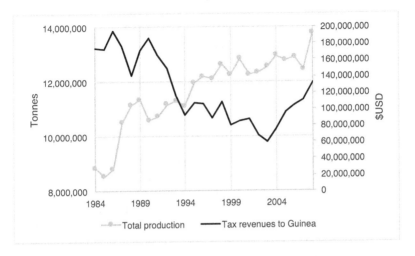

Figure 5.2 CBG's bauxite production and tax contribution to state revenues (1984–2008).

Source: Figure by author; data from CBG (2013).

making a direct contribution to regime stability. Mr Traoré, a retired employee of the mining sector and current consultant in mining and governance, suggests the following:

> Any regime in power knew that without any particular effort, the government had millions of dollars from mining taxations being paid to the state treasury from CBG alone. This amount enables the government to pay salaries and to pre-empt any sources of instability that could result from delay in the payment of civil servants salaries for instance.
>
> (interview, Traoré, 8 November 2013)

In addition to taxes paid to the state, CBG is the largest employer of workers in the industrial mining sector (CBG, 2013). Although CBG contributed to both employment and state revenues, the breakdown of its operating expenses shows that the company had high inequalities in terms of where its expenditure was allocated. In 2001, CBG employed 45 expatriate whose salaries accounted for USD$6.1 million, while the national payroll for its 2,435 employees was USD$17.3 million (MMG, 2005). In 2004, CBG employed 34 expatriate employees and 2,435 national employees, with payroll costs of USD$3.5 million and USD$18.7 million respectively (MMG, 2005; CBG, 2013). These numbers show that although CBG hired large numbers of Guineans, their salaries remained extremely low compared with those of its expatriate employees. Hence, it offered more benefit in terms of salaries and advantages to foreigners than it did for Guineans. During field interviews, the difference between national employees and expatriates was highlighted (interview, Anonymous 4, 7 November 2013).

Guinea has often been ranked amongst the most corrupt countries in the world; therefore, having a foreigner as a financial controller, for instance, makes it easier to manage funds and prevent frauds. Expatriates are brought in for skilled roles that cannot be filled locally or for strategic positions, such as financial controllers. According to an interviewee, "It is easier for an expat to say no to a minister or senior local authority than it will be for a Guinean" (interview, Anonymous 5, 7 November 2013).

Expatriates, even with minimum experience, were paid higher salaries than senior national staff. National employees felt that, despite, in some cases, having similar professional experience and international exposure to expatriate employees, the company still paid national employees relatively very little compared with what it paid expatriates. Regarding the remuneration of expatriates, most are paid salary relative to their country of origin, and in some cases their salaries are much higher than the average salary in Guinea because of the difference in cost of living. The average salaries in countries like Canada, Australia and the United Kingdom, for instance, are often much higher than salaries in Guinea.

The advantage for expatriates working in Guinea is that they receive benefits they do not receive at home, including housing, transport, security and paid holidays. On the other hand, if mining companies start paying national staff the same salaries as expatriates, the sector will create major inconsistencies within the Guinean economy. For instance, if national staff were paid as much as USD$50,000 or USD$100,00 per year, this would also exacerbate social instability, as they would earn 100 or 200 times the salary of a civil servant. Expatriates are often on temporary contracts, while national staff are offered permanent contracts; therefore, mining companies need to offer national staff salaries that are sustainable on the long term. Despite the perceived difference between national employees and expatriates, working at CBG remained a better option for Guineans compared with working in other sectors, including the civil service.

The majority of those who worked for mining companies had higher salaries than that of the average civil servant in Guinea; the average yearly salary of a national CBG employee was USD$5,000 per year, compared with USD$500 for civil servants before 2010 (Soumah, 2010). Although the amount paid to CBG national staff was ten times lower than the amount paid to expatriate staff, it was ten times higher than the average national wage (Soumah, 2010:45). Hence, many preferred to remain at CBG rather than work elsewhere in Guinea because, in light of employment opportunities, it was one of the best options in Guinea for highly skilled professionals. In addition to salaries, there were additional benefits to working for CBG, including free electricity, water, housing, health care, subsidised food rations, housing equipment and discounts on the purchase of vehicles (Basile, 2013). However, these advantages put further pressures on national employees of CBG, who often found themselves invaded by family members from across the country seeking a better education, health care and a chance to share in the welfare of families working on mining sites. Thus, the investment made by CBG in its employees had wider repercussions in terms of people's expectations and satisfaction or dissatisfaction with their situation.

5.7 Conclusion

Between 1958 and 2008, CBG was the largest bauxite mining company in Guinea, the highest contributor to state revenue, compared with other mining companies, and the highest employer outside the state. CBG made direct contributions to community development for people living in the mining areas and to the informal and the formal economy. Bauxite mining projects have led to the creation of new towns, new job opportunities and much-needed sources of revenue for Guinea. This chapter supports the argument, initially made in Chapter 3, which suggests that, at independence, Guinea's bauxite reserve was a key element in Guinea's development of bilateral commercial relationships with the Soviet Union, the United States and France, amongst others. The millions of dollars per year in tax revenues from bauxite mining were crucial to maintaining regime stability in Guinea. Between 1958 and 2008, bauxite mining benefited both the Touré and the Conté regimes and contributed to regime stability by providing both regimes with a secure and stable source of revenue.

CBG generated high revenues for the state in terms of taxes and offered employment opportunities to many Guineans. Despite these contributions, CBG and the state have been unable to meet the evolving expectations of local communities in terms of jobs and sufficient socio-economic development. This has resulted in organised protests by communities that have threatened CBG's mining activities. Although the state is a shareholder in CBG, it was CBG and not state institutions that provided solutions to communities' demands, and that is what calmed the populations protesting in the mining areas. Thus, CBG continued to be the intermediary between the state and citizens, which enabled the sustainability of the acquiescent permissive social contract that emerged.

Locally, bauxite mining companies have created enclaves that sustain a development gap between those who live inside the enclaves and those who do not. Because of the limited presence of the state in surrounding areas, CBG has, at times, been the main catalyst of social inequality and created social insecurity because of the gap between those living in CBG enclaves and those who do not, which has led to higher expectations and protests over the years. The standards created in the enclaves also attracted more people to the region of Boké, which put additional pressure on CBG and led to unmet expectations.

Although the state is a major shareholder of CBG, it has left all responsibility for socio-economic development to CBG, which created pressure on CBG to provide goods and services to communities in the mining region to ensure the functioning of their activities. The contribution of CBG to community development has resulted in the emergence of an informal acquiescent permissive social contract that CBG has been sustained on behalf of the state in its regions of activity. Today, with the growing number of bauxite mining companies in Guinea, we are also observing an informal acquiescent permissive social contract between mining companies and local communities. This will be briefly discussed in Chapter 6. In conclusion, the following statements can summarise discussions about CBG:

- CBG has been the main contributor to the socio-economic development of communities where it operates.
- Social insecurity has resulted in an increasing source of instability for the company. In this case, social insecurity has emerged because the government has left CBG to be the face of development, and it has disengaged from local communities in mining areas, leaving mining companies with a heavy burden of responsibility for providing employment opportunities and contributing to socio-economic development, which would reduce poverty in the region. Thus, expectations have risen amongst local communities *vis-à-vis* CBG.
- CBG was often obliged to find a quick solution to communities' concerns because it could not afford to stop production. The inability of CBG to meet production targets because of local protests results in less tax revenue for the country and less profit for the company, of which the government is a 49 per cent shareholder. The otherwise cash-poor government of Guinea could not afford to lose such a stable source of revenue, nor does the company want to lose money for its various shareholders. Hence, when CBG was targeted during protests, this was also an indirect attack on the government, with CBG acting as the intermediary needed to find a solution to the grievances of local communities.
- CBG was often left to create solutions to local rededications and protests. The importance of mining activities for both the company's and

the government's revenue meant that quick solutions had to be found to avoid prolonged periods of social insecurity, especially when the population threatens or resorts to protests.

• To ensure the long-term sustainability of the bauxite mining industry, there is an urgent need to understand how enclaves can be made to contribute to the wider local community and promote the kind of economic development that will sustain these mining towns long after mining has finished.

Between the years of this study (1958–2008), there was no formal contract between the state and mining companies that focused on the responsibilities of mining companies' contribution to socio-economic development in mining regions. Indeed, after 50 years of development of the extractive industry, Guinean society as a whole has yet to benefit significantly from the revenues of its mining sector. However, as discussed earlier, the state informally allowed CBG to perform state responsibilities in the mining regions. It has now become difficult for this to change because communities in mining regions are accustomed to direct support from CBG. It is not unusual for communities to expect mining companies to contribute to socio-economic development. However, in the case of Guinea, it has the largest reserves of bauxite in the world, which means that bauxite mining activities will continue for another 50 to 100 years. Therefore, to maintain stability it is vital for bauxite mining companies, together with the state, to find sustainable solutions to address the issues of socio-economic development in the mining regions.

Notes

1 Diallo (2017). The African towns falling into decline and poverty after mining companies use resources, then exit. Available from: https://theconversation.com/the-african-towns-falling-into-decline-and-poverty-after-mining-companies-use-resources-then-exit-69687
2 It was an interviewee's way to illustrate, the challenge the CBG faces- once it started investing in community development, local communities came to expect ongoing support which they later on felt entitled to.
3 Dr Doumbouya has been working in the hospital since 1984.
4 Before 1984, Guinea had a dictatorship and popular protests would have been banned. From 1984–1993, Guinea had a military regime. Although Conté is committed to civilian and democratic rule from 1993, the legacy dictatorship combined with military rule had built a strong fear of reprisal from state forces. However by 2004, the populace were no longer afraid to stand up to state forces and seek better socio-economic development.

References

Basile. B., 2013. *Avantages en nature*, Compagnie Des Bauxites de Guineé. November 2013.

108   *A case study of the CBG (1958–2008)*

Bond, P. 2007. Primitive Accumulation, Enclavity, Rural Marginalization and Articulation. In Centre for Civil Society pays tribute to Guy Mhone and José Negrão: Briefings' in the March 2007 *Review of African Political Economy*.

Boone, C., 2003. *Political Topographies of the African State: Territorial Authority and Institutional Choice*. Cambridge: Cambridge University Press.

Comité Technique de Revue des Titres et Conventions Miniers (CTRTCM), 2015. *Conventions Publiées*. [online] Republic of Guinea. Available from: www. contratsminiersguinee.org/about/projets.html [Accessed on 20 January 2015].

Compagnie des Bauxites de Guinée (CBG), 2008. *Quelques éléments de la Mission conjointe Rio-Tinto/CBG, de Diagnostic Social des Relations Sociocommunautaires de la CBG, du 1er Décembre au 05 Décembre 2008*. Kamsar: CBG.

Compagnie des Bauxites de Guinée (CBG), 2014. *Détails des Infrastructures Socio-Communautaires CBG (de 1987à 20014) Réalisès dans les Localitès Situèes dans la Zone d'Influence de CBG, Prefecture, Commune Urbaine, Sous-Prefectures et CRD; Kamsar-Kolaboui-Bintimodia-Boké-Sangarèdi-Missira- Daramagnaki et Institut Sup GèoMine Bokè, Tanene-Malapouya- Sansalé- Kanfarandé-Dabiss, 2014*. Data Accessed with permission of CBG, June 2014. Kamsar: CBG.

Compagnie des Bauxites de Guineé (CBG), 2019. *Liste des TPE lancees de 2010–2017 a travers le fond revolving CBG*. Document de CBG.

Compagnie des Bauxites de Guinee by Touré, K., 2013. *Compagnie des Bauxites de Guineé (CBG), 2013-A Guinean Success Story, June 2013*. London: CBG.

Diallo, P., 2017. Social Insecurity, Stability and the Politics in West Africa: A Case Study of Artisanal and Small-Scale Diamond Mining in Guinea, 1958–2008. *The Extractive Industries and Society*. 4(3), pp. 489–496. DOI:10.1016/j.exis.2017.04.003.

Ferguson, J., 2005. Seeing Like an Oil Company: Space, Security, and Global Capital in Neoliberal Africa. *American Anthropologist*. 107(3), pp. 377–382.

Ferguson, J., 2006. *Global Shadows: Africa in the Neoliberal World Order*. Durham, NC: Duke University Press.

Le Vine,V. T., 2004. *Politics in Francophone Africa*. Boulder: Lynne Rienner.

Ministere des Mines et de la Geologie (MMG), 2004. *Direction des Etudes et de la Prospective. Memorandum, Objet: RUSSKY Aluminy (Rusal) en Guinée*. Ministere des Mines et de la Geologie, Direction des Etudes et de la Prospective, Republique de Guinée, Conakry.

Ministry of Mines and Geology (MMG), 2005. *Guinea, Mineral Resources, Bauxite*. Republic of Guinea. Conakry: MMG.

Olukoshi, A., 2006. *Enclavity: A Tribute to Guy Mhone*, Keynote address to the Colloquium on Economy, Society and Nature, University of KwaZulu-Natal Centre for Civil Society, Durban, 1 March. In CCS pays tribute to Guy Mhone and José Negrão: Briefings' in the March 2007. *Review of African Political Economy*, 'Primitive accumulation, enclavity, rural marginalization and articulation' By Patrick Bond.

Snyder, R., Bhavnani, R., 2005. Diamonds, Blood, and Taxes a Revenue-Centered Framework for Explaining Political Order. *Journal of Conflict Resolution*. 49(4), 563–597.

Soumah, I., 2008. *The Future of Mining Industry in Guinea*. Paris: L'Harmattan.

Soumah, I., 2010. *Les Mines de Guinée: Comment Cela Fonctionne*. L'harmattan, Paris.

Touré, K., 2013. *Compagnie des Bauxites de Guinée (CBG), 2013-A Guinean Success Story, June 2013*. London: CBG.

6 Conclusion

We have looked at how the availability of bauxite resources facilitated the coexistence of regime stability and social insecurity in Guinea during the period 1958–2008. It is hoped that this book contributes to and will encourage more studies on the stabilising and destabilising potential of mining, as opposed to the one-sided focus of most recent work in this field, which typically focuses either on the negative aspect of lootable resources (Snyder, 2006; Snyder & Bhavnani, 2005; Bannon & Collier, 2003) or on the contribution of non-lootable resources to regime stability (Dunning, 2008; Snyder & Bhavnani, 2005).

The main argument is that the Touré (1958–1984) and the Conté (1984–2008) regimes in Guinea both maintained regime stability in the face of social insecurity by using mineral resources to strengthen their hold; this, in turn, facilitated the emergence of different forms of social contracts in Guinea's post-independence era. Bauxite mining catered to the revenue needs of the regimes by providing them with a stable source of revenue. In addition, bauxite mining companies such as CBG made extensive contributions to local community development, which in later years addressed and largely contained local conflict in mining areas. Although this balance was fragile, it enabled Guinean regimes to maintain the status quo for the first 50 years of the country's independence.

6.1 Carving a new path – Guinea, mineral extraction, social contracts and stability

Within academic literature on mineral extraction in West Africa, very little attention has been given to Guinea. The country deserves scrutiny because it is an exceptional case, having avoided mineral resource–linked large-scale conflict despite the dictatorial nature of its regimes, shortcomings in governance and high poverty rates. This book investigates the hitherto unexplored topic of how mineral resources can contribute to the coexistence of regime stability and social insecurity, the successful navigation of which enables regimes to avoid conflict and maintain the status quo. By studying Guinea, this book has introduced a different lens through which to examine resource-rich countries. Looking further at how Guinea fails to conform to previous academic theorising about the state and mineral resources and what can be learnt from this, this book has contributed to the current literature in four ways.

First, arguments put forward regarding the so-called resource curse do not always happen as predicted, and a country such as Guinea can have all the symptoms of the resource curse without descending into large-scale conflict. In contrast to Sierra Leone, for instance, Guinea was able to escape large-scale mineral resource–linked conflict despite the country's shortcomings in governance.

Second, the successful failed state (Soares de Oliveira, 2007) and the rentier state (Yates, 2009) concepts focus on oil extraction to explain its contributions to regime stability. We see here that bauxite extraction operates in comparable ways to those of oil extraction. The impact of bauxite mining is comparable to that of oil extraction. In the case Guinea, bauxite mining is the largest industry in the country, and it is a major contributor to states revenues, thus enabling the state to neglect other development sectors such as agriculture or infrastructure. The presence of large bauxite mining companies such as CBG offers the state guaranteed access to sustainable sources of revenue. From CBG alone, the state had access to millions of dollars in taxes, and this played a key role in contributing to the economic stability of the regimes. Like the oil industry, it has built enclaves that have become a catalyst for grievances towards mining companies. Lastly, just like oil rich countries, such as those in the gulf of Guinea, revenues from bauxite mining largely benefit the state and mining companies.

Third, previous studies by Snyder and Bhavnani (2005) offer an insight into how non-lootable resources contribute to regime stability. The unique aspect of this study is that it focuses on how non-lootable bauxite mining resources contribute to regime stability by sustaining new forms of social contracts. This clearly extends the work of Snyder and Bhavnani (2005), setting it in a wider context. This should be taken further in studies of other countries and contexts.

Finally, the thesis in this book has used the social contract as a theoretical framework. However, in addition to Nugent's (2010) four types of social contract, namely coercive, productive, liberational and permissive contracts, this book introduces four additional typologies, namely strong coercive social contract, limited coercive social contract, acquiescent permissive social contract and circumstantial permissive social contract. This extension of the existing academic theoretical framework was found necessary because it offered the opportunity to take into account further underlying dynamics related to state–society relationships in a country like Guinea that depends highly on mineral resource extraction. Hence, these typologies offer a detailed view on state–society relations in the African context, while the book offers an opportunity to take into account dynamics that have not previously been accounted for in the literature on state–society relations in Africa.

6.2 Findings

6.2.1 *Regime stability: maintaining a delicate balance*

Regime stability was maintained in Guinea from 1958 to 2008 through the emergence of social contracts that enabled regimes to maintain coercive tools to control citizens and manipulate domestic politics. These social contracts also, in turn, enabled rural societies to function independently of the state in bauxite mining regions. During this period, rural societies presented no major challenge to regime stability. However, in cases where the state restricted or delayed access of citizens to socio-economic services such as water and electricity, the ensuing frustration combined with hardship and lack of employment created the potential for instability. This led to some clashes between CBG and local communities. Typically, these were resolved through negotiations and provisions of further community development projects before they could erupt into large-scale conflict.

Touré was able to maintain the coexistence of regime stability and social insecurity for 26 years because he was able to maintain a strong coercive social contract between the state and citizens. Bauxite was the main enabling asset, providing a significant and stable source of revenue that did not largely depend on taxing.

The ways in which the Touré (1958–1984) and the Conté (1984–2008) regimes maintained their rule were different, despite bauxite being the main enabling asset for each. Guinea's bauxite wealth contributed to Touré's ability to build key political and commercial relationships with the Soviet Union, the latter becoming Guinea's main political supporter and business partner. In addition to the Soviet Union, mining enabled Guinea to build a strong economic partnership with countries such as the United States, which enabled the Touré regime to gain external legitimacy despite France's attempt to sabotage Guinea's independence. Mining enabled Touré to enact his nationalistic, dictatorial and anti-imperialist views. Without mineral resources, it is unlikely that Guinea would have survived without France's support, because the country was poor and had no economic, financial or human resource capacities to move forward. Soviet interest in Guinea's bauxite made it possible to replace French support with Soviet support.

The second regime, led by Conté, demonstrated that it is possible to maintain the coexistence of regime stability and social insecurity with a less coercive social contract between state and citizens than that adopted by Touré. In this case, regime stability was maintained by Conté's ability to use mining resources to maintain loyalty from the army. More importantly, when Guinea faced the threat from Sierra Leonean and Liberian rebels, Conté was able to use mining assets to fight back the rebellion and avoid a slide into conflict. Although Conté had the loyalty of the majority of the army, without access to mineral resource assets he would have not had the financial resources to fight back the rebellion.

Although the Conté regime remained dictatorial, the country was opened to transnational organisations and to monitoring by international institutions such as the World Bank and the IMF. Although Guinea became formally democratic and held multi-party elections, these changes did not improve socio-economic conditions. In fact, the country remained amongst the poorest in the world, and poverty increased from 2004 to 2008 as the country experienced high rates of inflation. Despite these socio-economic issues, Guinea managed to avoid large-scale conflict.

6.2.2   *Bauxite mining-revenues, contributions to state and society and the social contract*

Crucially, the reliance of Touré and Conté on the mining sector created a situation in which the state in Guinea became detached from and unaccountable to its citizens.

The presence of large industrial bauxite mining companies such as CBG offers the state guaranteed access to a sustainable source of revenue. From CBG alone, the state had access to millions of US dollars in taxes, and this played a key role in contributing to the economic stability of the regimes. The case of CBG also shows that in areas where industrial mining of bauxite takes place, the state was often absent. The failure of the state to provide goods and services to communities in the bauxite mining region created pressure on CBG to provide goods and services to communities, including electricity, water, markets, health centres and schools. Ultimately, this rendered the state unaccountable to its citizens. The state informally allowed CBG to become responsible for those communities. This acquiescent permissive social contract between the state and citizens depended on CBG as an intermediary agency to deliver goods and services.

This disengagement of the government *vis-à-vis* communities in mining areas and the delivery of socio-economic goods and services by CBG in turn increased communities' expectations of CBG. As a result, CBG became vulnerable when communities' expectations were not met. When communities wanted to gain the attention of the government regarding their socio-economic conditions, they resorted to protests threatening the activities of CBG. In return, the government often left it to CBG to respond.

Revenues from the mining industry are yet to contribute to sustainable development in Guinea for the general population. Despite disappointing results and increasing poverty, poor socio-economic conditions and the regimes' inability to provide Guinean society with basic socio-economic goods, Guinea had, by 2008, survived 50 years of dictatorial regimes without large-scale mineral related conflict.

6.3   Future outlook for mining, social insecurity, social contracts and politics in Guinea from 2010 to present

Guinea democratically elected its first president in 2010, and the reform of the mining sector was one of the first pledges of President Alpha Condé. As the population becomes aware of the contributions made by mining companies and the responsibilities of government to its constituencies, governments face further demands from their citizens, as was seen during the protests in 2006–2008. In the future, governments and private industrial mining companies will need to ensure that local communities receive more benefit from mineral extraction revenues and employment opportunities. Whether more effective impact-benefit agreements between companies and communities, such as those in Canada and Australia, or between the state and citizenry can be made to function – and how this repositions multinational companies – remains to be seen.

Since 2010, although there has been an increase in the presence of bauxite mining companies in Guinea, the industry has yet to contribute to wider socio-economic development within the mining region or across Guinea. This has led to increasing instability and youth protests across Boké. A major threat to mining, politics and stability in Guinea lies in the presence of bauxite wealth, coupled with an unprecedented flow of mining companies, in the midst of poverty and growing youth unemployment. The consequence of these issues are discussed in the following sections.

6.3.1   *National wealth, personal poverty: living in the "paradox of plenty"*

We the citizens of Boké are convinced that for the steady evolution of mining activities there needs to be a principle of good neighborhood between mining companies, local populations, and the state. To achieve stability, mining companies and the state need to profoundly change their attitude and create a new partnership based on respect and the application of the changes that we have requested.

(*Mémorandum Des Population de Boké*, 2017:6)[1]

The youth are disappointed. . . . We know the government is unhappy because they know that we are the ones who can affect their earnings during protests. We want water, electricity and youth employment otherwise the mining companies will leave! We are ready to fight for what we feel are our rights! . . . We are not against the mining activities. However, we want to benefit from the mining activities. Otherwise, we would rather the mining companies stop mining.

(focus group discussion, 26 April 2018, Boké)

Despite the presence of bauxite mining companies, local communities face high unemployment, poverty and a lack of adequate socio-economic and essential services, such as water and electricity (*Mémorandum Des Population de Boké*, 2017). With poor governance and corruption, local youths are finding jobs in mining companies inaccessible. In addition to dealing with the environmental impact of bauxite mining, communities have to deal with the increasing cost of living because of mining activities. Thus, the life of local communities and particularly youths in the region of Boké has become a daily struggle in the midst of mineral wealth. Youths are no longer content with watching bauxite trucks moving out of the region to feed a global trade while the majority are struggling to make ends meet.

Protest in the region of Boké is not new. However, in recent years, the protests are being led by young people determined to see a positive change. To express their anger, youth protestors often block transport routes and destroy public buildings to gain the attention of the state and mining companies.

With the increasing number of mining companies, communities' demands are no longer focused on socio-economic growth – they are also geared towards environmental concerns and inclusion. The new protests are a result of the changing landscape created by mining activities. Communities feel that they have become observers collecting negative environmental impacts instead of active stakeholders benefiting positively from the effects of mining activities. The government presence on the ground is limited, and this has facilitated the increasing grievance between local communities and mining companies.

6.3.1.1   *CBG: a model of public private partnership for service delivery?*

The first company to be operational in the bauxite regions of Boké was CBG. Over the past 30 years, CBG has made significant and evolving forms of contributions to local communities and its employees. As a result, new mining companies are expected to provide similar services as those offered by CBG. However, while CBG built hospitals and roads that are accessible to the public and the community, some of the new companies are using public roads to transport bauxite and have not invested in new infrastructure that will benefit the local community. Instead, these new companies often place pressure on existing infrastructure and create further tension amongst communities that also rely on these public assets.

The high expectations for new mining companies to follow the model of CBG regarding the provision of services constitute a major challenge. According to Keita (interview, Boké, 27 April 2018), "the evolution of the CBG provided different social services to communities including health services, water, and electricity. Communities now think that mining companies' arrival is synonym with social service". This foreshadows that bauxite mining will continue as a source of tension in the region of Boké with ongoing clashes between local claims and international and state interests.

For the most part, marginalised African youths only attract global attention when conflict occurs, as in the case of Sierra Leone and Liberia (Hoffman, 2006; Richards, 2006; Kandeh, 1999; Abdullah, 1998; Bangura, 1997; Rashid, 1997). In Sierra Leone, the presence of unemployed youths coupled with real grievances can make them an asset for organisd rebellions. Some authors have linked conflict to the youth crisis in Sierra Leone, which has been linked to the state's failure to provide adequate socio-economic development and employment opportunities for them (Hoffman, 2006; Richards, 2006; Kandeh, 1999; Abdullah, 1998; Bangura, 1997; Rashid, 1997). These authors argue that it was easy to manipulate youths to join the rebellion in Sierra Leone as most of them in rural areas were unemployed, uneducated, marginalised and excluded from the government elites (Gberie, 2005).

By contrast, Guinea has historically not figured as such a site of armed conflict. And the agency of young people has been found to be far from negative. In general, "young people in Guinea were key to maintaining social cohesion in a politically-volatile national and regional context" (Wybrow & Diallo, 2009:1).

However Guinea is now seeing a wave of youth-led protests demanding improved socio-economic conditions, including improved education systems all over the country.[2] Guinea is facing a youth crisis that, if not treated, will lead to violent conflict. Recent events in Guinea are especially worrying. The case of Boké shows that young people are no longer passive actors in relation to changes in their environment. Exclusion from mining activities has increased mistrust towards mining companies, NGOs, the state and its representatives. Young people are no longer afraid to disturb social cohesion to seek change and improved outcomes.

Government institutions are still not effectively addressing some of the issues and mounting grievances of youth concerns. No mining company, even CBG, can unilaterally address the issue of youth unemployment in the mining region and in the region of Boké. It is crucial for the state to start looking for a practical, long-term solution to youth issues in Boké, where youths are determined to voice their concerns until satisfied and are no longer afraid to use violence to this end. As seen in the case of Sierra Leone and Liberia, the consequences of violent protests are unpredictable. The role played by mineral resources in the civil wars in Liberia and Sierra Leone were fundamental to conflict, presenting immediate threats to people, regimes and businesses. Disregarding the current situation of youths in Boké will lead to further violence and unwanted consequences for mining companies and the state.

To maintain stability in the bauxite mining region of Boké, new companies will have no choice – because of limited contribution of the state in the socio-economic development of bauxite mining regions and the reliance of communities on industrial mining companies – but to sustain acquiescent permissive social contracts, which emerged with CBG. The next section, which is the end of this chapter, offers a reflection on the African Mining Vision.

6.4   Reflection on the African Mining Vision (AMV)

> The Africa Mining Vision envisages "transparent, equitable and optimal exploitation of mineral resources to underpin broad-based sustainable growth and socio-economic development".
>
> (African Union, 2009:5)

The Africa Mining Vision (AMV) was launched by the African Union in 2008 and adopted by the African heads of states and governments in February 2009. The AMV is "Africa's own response to tackling the paradox of great mineral wealth existing side by side with pervasive poverty" (UNECA, n.d.). The adoption of the AMV is a major initiative for Africa by African leaders: this is the first time that African leaders rather than external actors had developed an initiative to improve the conduct of the mining sector.

Africa is endowed with one of the world's largest reserves of minerals, including bauxite, diamonds, iron ore, gold, platinum, chromite, manganese and vanadium (Bush, 2008). From 1986 to the 1990s, in an attempt to increase the contribution of the mining sector to socio-economic development, many economic and political reforms and initiatives were implemented in resource-rich African countries.

From the 1980s to 1990s, the World Bank (WB) and the IMF (International Monetary Fund) led the implementation of several reforms and policies that were meant to attract Foreign Direct Investments (FDI) and contribute to socio-economic development (Bush, 2008; Campbell, 2004). These reforms led to the liberalisation of economies and the privatisation of the mining industry across Africa. However, as illustrated in Chapter 5 the results and consequences of reforms led by the World Bank and the IMF were disappointing. The liberalisation of the African economy in the 1980s and 1990s did increase FDI, especially in resource-rich countries, but while it increased the production of most minerals and the profits of mining companies, it did not result in the predicted socio-economic development of local populations (Campbell, 2004; Graham, 2013).

In general, the liberalisation of the mining sector benefited only private mining companies, which took advantage of reduced taxes, favourable policies, easier access to the African mining sector and a lack of control by governments, which were relegated to a purely administrative role, resulting in the private mining companies being unaccountable for their activities (Bush and Graham, 2010; Campbell, 2004).

The minimising of state involvement in the mining sector meant there was no space for the reforms to either build the state's institutional capacity or develop state involvement in their effective implementation (Campbell, 2004). Additionally, the benefit of mining to local communities was limited to a few examples of developments of the informal economy in areas around mining enclaves. Nonetheless, the mining industry remains the major sector from which most resource-rich African countries hope to promote socio-economic development.

6.4.1 The Africa Mining Vision – a panacea to the challenges of the African mining sector or another mirage?

The AMV takes a holistic approach, integrating mining into national policies and ensuring that mining contributes to sustainable development (UNECA, 2014). It promotes small-scale mining, better environmental policies, institutional and human capacity building, good governance, transparency and accountability, with the aim of adding value to mineral resources before exporting them and developing economic linkages (UNECA, 2008; AU, 2009; ACEP, 2014). Thus, it prioritises local development rather than attracting FDIs, as was the case with previous reforms undertaken by the World Bank and the IMF.

Today, ten years after adopting the AMV, despite its objective of transforming the African mining sector, not much has changed. There are unresolved problems relating to artisanal and small-scale mining, and there is an urgent need to build the human and institutional capacities of state institutions (Bush, 2008:4–7). The challenges still facing the African mining sector include transparency and accountability; governance and public participation; the environmental, economic, social and health impacts of mining operations; and the sector's failure to add value or provide benefits relating to research, development and technology.

While an increasing number of industrial mining companies have been operating across the continent, socio-economic growth for the larger population has yet to follow. Countries rich in natural resources, including Guinea, Niger, Sierra Leone, Nigeria, Equatorial Guinea and Chad, still have a majority of their populations living in extreme poverty with high rates of unemployment. These countries still suffer from corruption that limits the ability of most citizens to benefit from mining revenues. In Guinea, for instance, the majority of the population lives in poverty, and much of the population of the country still lacks access to clean drinking water, health centres, electricity and roads.

From 2008 to 2015, efforts seem to have been invested in measures that would support the effective implementation of the AMV. One area was the creation of the African Mineral Development Centre (AMDC) in 2011. The AMDC published a guidebook in 2014 to ensure that the objectives of the AMV would be effectively promoted at the country level and that member states would align their policies and objectives to those of the AMV (UNECA, 2014). In addition, there have been several efforts in putting information relating to mining policies of different African countries online. Although these are commendable efforts, ten years would seem to represent a long time to wait for a vision to be implemented. It is now time for the AMV to move to the implementation phase before the momentum that led to its creation is lost.

Previous mining initiatives have not succeeded in effectively ensuring that mining revenues are linked to the wider local economy (Bush, 2008). Many issues need to be addressed to ensure that the AMV is a successful solution to the challenges presented by the African mining industry. In practice, socio-economic development resulting from the activities of the mining industry cannot be accomplished without building the resource-rich states' institutional capacity to implement, promote and monitor reforms within the mining industry. Additionally, successful mining activities can only be achieved if local communities feel that they would benefit from mining projects.

Based on my observations from the field and the literature I have surveyed, in the majority of African countries, the most urgent need is to increase the human and institutional capacity of country-level stakeholders to implement, monitor and address the issues associated with mineral resource extraction. I see the following four points as crucial for the AMV in addressing the capacity-building challenges across the continent and thus becoming the solution that it hopes to be rather than a mirage of empty promises.

The first is to effectively build the capacity of governments to design and enforce viable environmental and development-sensitive mining policies. Perhaps in part because of mining's contribution to employment and to the informal economy, the environmental impact of mining remains a major challenge that has yet to be adequately addressed in most African countries (Bush, 2008). Although most African countries have environmental legislation, this is often not enforced. In most countries, including Guinea, Sierra Leone and Liberia, mining has posed real development and environmental challenges. Agricultural lands are used as mining sites, and once mining is finished, there is no effort to make the land viable for agriculture once again. Capacity-building ideas could include helping governments design and implement viable, environmentally sensitive mining policies. These policies could, for instance, include the transformation of post-mining sites to viable agricultural land, which would provide sustainable livelihoods for rural communities. In this way, livelihoods would be sustained even after mining activities have ceased. Thus, African governments need to develop the skills, legislation, culture and will to regulate mining companies working on the continent and ensure that penalties are paid by those polluting the environment or putting the health of local citizens at risk.

Second, African governments need to develop their ability to capture artisanal mining revenue in the formal economy; many are currently unable to do so. The additional revenues from the artisanal mining sector would give governments further incentive to focus on artisanal mining communities and support their socio-economic development.

Third, local NGOs need to develop their ability to campaign and advocate for better mining policies and operations and to alert and educate communities about environmentally sensitive approaches to mining. Stronger and more capable grassroots NGOs would be able to hold governments to account for their policies and advocate policies and practices for sustainable economic development and environmentally sensitive mining.

Fourth, the FPIC (Free, Prior and Informed Consent)[3] principle should be respected in all countries during the development of mining projects. FPIC

> is a specific right that pertains to Indigenous Peoples and is recognized in the UNDRIP. It allows them to give or withhold consent to a project that may affect them or their territories. Once they have given their consent, they can withdraw it at any stage. Furthermore, FPIC enables them to negotiate the conditions under which the project will be designed, implemented, monitored and evaluated.
>
> (FAO, 2016:13)

Mining has a direct impact on communities and their environment. Therefore, all mining projects should be developed in line with FPIC principles. Indigenous communities around the world have different stories, realities and decision-making approaches; therefore, it is difficult to adopt one fit for all forms of "a one-size-fits-all formulation of FPIC" (ASI, 2017:69). Mining projects need to take into account the differences amongst communities and ensure that the FPIC principles are implemented in line with local approaches.

Mining companies and governments are facing ongoing pressure from communities demanding to be engaged in decisions regarding their land. This means both governments and mining companies need to ensure that local communities are involved in and approve decisions affecting their lands to avoid any disruption to mining activities once they start. The adequate implementation of FPIC principles could therefore enable a more stable relationship amongst different stakeholders.

If the AMV continues to take longer in delivering on its vision, we will see more community-led protests on mining sites, creating further instability in the mining industry, less profit for private companies and lower revenues for mineral resource–rich states. Mining will only be successful where stability of production can be ensured, but this can only be guaranteed with the support of the local communities. With the Internet and the growing role of NGOs in the African mining sector, local communities know that they no longer have to be passive victims; rather, they can be actors bringing change to their communities. So far, this has only been achieved through protests at mining sites. It would be much more positive and productive if measures were in place to ensure that the benefits of mining activities are shared beyond mining enclaves.

The success of the AMV will depend on the ability of African leaders to ensure that gaps in human resources and institutional capacity in specific countries are filled. Populations in mineral resource–rich countries are becoming impatient and frustrated, and the AMV needs to meet their legitimate socio-economic expectations that mining revenues should benefit local communities and contribute to socio-economic development in their country as a whole.

To prevent further social instability generated by the extractive industry in Africa, the AMV needs to make its mark – and make it very soon.

Notes

1 The Boké Population Memorandum was signed on 27 April 2017 by the president of the Communal Council of Youth, the women's representative, the Religious Confessions, the president of the Prefectural Council of Civil Society Organizations, the representatives of the District Councils, the Countigui de Boké, the president of the Prefectural Council of Civil Society Organizations and the president of the local district. This document includes the list of concerns, expectations

and demands from the local communities in return for the stable extraction of bauxite in the region of Boké.

2 In May 2018, when I arrived in Guinea, I saw two electricity distribution infrastructures that were burned by young protesters – one in the capital, Conakry, and the other in the town of Boké. This shows equal dissatisfaction of youths in both mining and non-mining areas, which raises concerns of a broader youth crisis in Guinea. However, this book does not intend to explore issues beyond Boké.

3 For definition of the principles please see "ASI indigenous peoples advisory forum in ASI performance standard V2: Guidance", December 2017, p. 69, available at www.aluminium-stewardship.org.

References

Abdullah, I., 1998. Bush path to Destruction: The Origin and Character of the Revolutionary United Front, Sierra Leone. *Journal of Modern African Studies*. 36(2), pp. 203–235.

Africa Centre for Energy Policy (ACEP), 2014. *A Guide to the Africa Mining Vision (AMV)*. Accra: ACEP.

African Union (AU), 2009a. *Africa Mining Vision*. Addis Ababa: AU.

African Union (AU), 2009b. *Africa Mining Vision: African Union*. Available from: http://africaminingvision.org/amv_resources/AMV/Africa_Mining_Vision_English. pdf [Accessed on 24 July 2015].

Aluminium Stewardship Initiative (ASI), 2017. ASI Indigenous Peoples Advisory Forum in ASI Performance Standard V2: Guidance, December 2017, p. 69. Available from: www.aluminium-stewardship.org [Accessed on 24 July 2019].

Bangura, Y., 1997. Understanding the Political and Cultural Dynamics of the Sierra Leone War: A Critique of Paul Richards's Fighting for the Rainforest. *African Development*. 22(3/4), pp. 117–148.

Bannon, I. and Collier, P., 2003. Natural Resources and Conflict: What We Can Do. In *Natural Resources and Violent Conflict: Options and Actions*. Washington, DC: World Bank.

Bush, R.C., 2008. Soon There Will Be No-One Left to Take the Corpses to the Morgue: Accumulation and Abjection in Ghana's Mining Communities. *Resources Policy*. 34(1–2), pp. 57–62.

Bush, R. & Graham. Y., 2010. *Mining Companies Are Not Interested in Africa's Development*. The Guardian. Available from: http://www.theguardian.com/business/2010/dec/02/eu-raw-materials-initiative-africa-mining [Accessed on 29 December 2015].

Campbell, B., (ed.), 2004. *Regulating Mining in Africa: For Whose Benefit?* Discussion Paper No. 26. Uppsala: Nordiska Afrikainstitutet.

Dunning, T., 2008. *Crude Democracy: Natural Resource Wealth and Political Regimes*. New York: Cambridge University Press.

Food and Agriculture Organization of the United Nations (FAO), 2016. *Free Prior and Informed Consent: An Indigenous Peoples' Right and a Good Practice for Local Communities*. United Nations Department of Economic and Social

Affairs Indigenous Peoples, p. 16. Available from: www.fao.org/3/a-i6190e.pdf [Accessed on 29 December 2018].

Gberie, L., 2005. *Dirty War in West Africa: The RUF and the Destruction of Sierra Leone.* London: Hurst and Co.

Graham, Y., 2013. *Escaping the Winner's Curse: The Africa Mining Vision and the Challenges of the International Trade and Investment Regime.* Workshop on International Law, Natural Resources and Sustainable Development, 11–13 September, 2013, Third World Network Africa, Accra.

Hoffman, D., 2006. Disagreement: Dissent Politics and the War in Sierra Leone. *Africa Today.* 52(3), p. 3–22.

Kandeh, J., 1999. Ransoming the State: Elite Origins of Subaltern Terror in Sierra Leone. *Review of African Political Economy,* 26(81), pp. 349–366.

Mémorandum Des Population de Boké, 2017. p6 signe par le Président du Conseil Communal de la jeunesse; La Représentante des femmes, Les confessions Religieuses, Le Président du Conseil Préfectoral des Organisations de la Société Civile, Les représentants des Conseils de Quartier, Le Countigui de Boké, Le Président du Conseil Préfectoral des Organisations de la Société Civile, Le Président du CPD. Signe a Boké le 27 Avril, 2017.

Nugent, P., 2010.States and Social Contracts in Africa. *New Left Review.* 63, pp. 35–68.

Oliveira, R.M.S.D., 2007. *Oil and Politics in the Gulf of Guinea.* London: C. Hurst & Co.

Rashid, I., 1997. Subaltern Reactions: Student Radicals and Lumpen Youth in Sierra Leone, 1977–1992. *African Development.* 22(3/4), 19–43.

Richards, P., 2006. Young Men and Gender in War and Post-war Reconstruction: Some Comparative Findings from Liberia and Sierra Leone. In: Bannon, I. and Correia, M., ed. *The Other Half of Gender: Men's Issues in Development.* Washington, DC: World Bank, pp. 195–218.

Snyder, R., 2006. Does Lootable Wealth Breed Disorder? A Political Economy of Extraction Framework. *Comparative Political Studies.* 39(8), pp. 943–968.

Snyder, R. and Bhavnani, R., 2005. Diamonds, Blood, and Taxes a Revenue-Centered Framework for Explaining Political Order. *Journal of Conflict Resolution.* 49(4), 563–597.

United Nations Economic Commission for Africa (UNECA), 2014. *Minerals Centre Produces Guidebook for Domestication of African Mining Vision.* Addis Ababa: UNECA.

United Nations, Economic Commission for Africa (UNECA), n.d. *About the Africa Mining Vision.* Addis Ababa: UNECA.

The United Nations Economic Commission for Africa (UNECA), 2008. *Africa Review Report on Mining -Executive Summary.* Addis Ababa: UNECA.

Wybrow, D. and Diallo, P., 2009. *Youth Vulnerability and Exclusion (YOVEX) in Guinea: Key Research Findings.* London: Conflict, Security and Development Group Papers Summary, King's College London.

Yates, D., 2009. Enhancing the Governance of Africa's Oil Sector, *SAIIA Occasional Paper* 51, Johannesburg: SAIIA, South African Institute of International Affairs.

Index